A THEORY OF

Global Biodiversity

MONOGRAPHS IN POPULATION BIOLOGY

SIMON A. LEVIN AND HENRY S. HORN, SERIES EDITORS

A complete series list follows the index.

A THEORY OF
Global Biodiversity

BORIS WORM

AND

DEREK P. TITTENSOR

PRINCETON UNIVERSITY PRESS
Princeton and Oxford

Published by Princeton University Press,
41 William Street, Princeton, New Jersey 08540

In the United Kingdom: Princeton University Press,
6 Oxford Street, Woodstock, Oxfordshire OX20 1TR

press.princeton.edu

Library of Congress Control Number 2017958993
ISBN 978-0-691-15483-1

British Library Cataloging-in-Publication Data is available

This book has been composed in Times Roman

Printed on acid-free paper. ∞

Printed in the United States of America

10 9 8 7 6 5 4 3 2 1

This work is dedicated to Sylvie Moe, and all the other children

who will inherit the fragile beauty that graces our planet.

Contents

Acknowledgments

This volume attempts an empirical and theoretical synthesis of global patterns in biodiversity. Clearly, any synthetic work in this field, including this one, first and foremost builds on the detailed taxonomic and biogeographical surveys performed by thousands of dedicated and inspired colleagues. We admire, and feel greatly indebted to, these collective efforts. Without the venerable and frequently unrecognized persistence of naturalists, taxonomists, field researchers, evolutionary scientists, and paleontologists, both past and present, none of this work could have been attempted. Furthermore, any conceptual advances herein could not have emerged without the foundational work on spatial patterns in biodiversity by many others. We feel especially lucky to have been inspired by discussions we had with some of the great minds in the field. These include our former mentors Anthony R. O. Chapman, Ransom A. Myers, and Ulrich Sommer, who helped to shape our scientific inquiry into nature's patterns and processes. We have greatly benefited from and are deeply grateful to our friends, colleagues, and students Drew Allen, Greg Britten, Jim Brown, James Estes, John Grady, John Grant, Mike Harfoot, Stephen Hubbell, Walter Jetz, Mark Johnston, Heike Lotze, Eric Mills, Camilo Mora, Tim Newbold, Robert Paine, Drew Purves, Thorsten Reusch, Gabriel Reygondeau, James Rosindell, Edward Vanden Berghe, Hal Whitehead, and many others who have provided much valuable feedback, critical advice and inspiring debate and discussion over the years.

Special thanks also to colleagues and institutions that shared data or made them freely available on the Web for the purposes of this volume. These include Kevin Gaston, Benjamin Halpern, David Jablonski, Kristin Kaschner, Stuart Kininmonth, Holger Kreft, Jorge Molinos, Willem Renema, Callum Roberts, Jim Valentine, Skip Woolley, NOAA, and the International Union for Conservation of Nature (IUCN). We gratefully acknowledge further support by colleagues and staff at Dalhousie University in Halifax. We also thank Dalhousie University, the UN Environment World Conservation Monitoring Centre, and Neil Burgess, Stephen Emmott, Jon Hutton, Tim Johnson, Sally Newton, and Jörn Scharlemann for providing an environment and opportunity conducive to making this undertaking possible. We also thank the Census of Marine Life and the Natural Science and Engineering Research Council of Canada for financial support. We feel indebted to our partners, Heike Lotze and Andrea Moe, who were incredibly supportive over the period of writing this volume. Finally, we are grateful to the Trident coffee shop and The Henry House, Halifax, for providing inspiring writing environments and appropriate hydration.

A Theory of
Global Biodiversity

CHAPTER ONE

Introduction

The most striking feature of Earth is the existence of life,
and the most striking feature of life is its diversity.
—David Tilman

Our planet is characterized by two highly unusual, and intriguingly beautiful, aspects: the fact that it contains life at all, and the spectacular ways in which novel life forms have diversified and assembled into complex communities and ecosystems. We have not—at the time of writing—discovered life on other celestial bodies, and even if life forms do exist elsewhere, the complexity of Earth's diversity may still remain unique or inordinately rare. As far as we know, the first microbial life on Earth originated in a primordial abiotic ocean some 3 to 4 billion years ago, possibly at hydrothermal vents (Martin et al. 2008), although several competing hypotheses exist (Mulkidjanian et al. 2012). From there, the remarkable biodiversity that now defines and shapes life on our planet evolved further, dispersing across the global ocean, and, about 0.6 billion years ago, onto the land (Retallack 2013). A cornucopia of profound biochemical, physiological, structural, and behavioral developments and adaptations facilitated this expansion into previously lifeless and often hostile environments. The evolutionary history of marine to terrestrial colonization is still visible in the distribution of higher taxa today: almost all animal phyla occur in the oceans, while fewer than half exist on land (May 1988). Yet, total species richness of eukaryotes is estimated to be about threefold greater on land (Mora et al. 2011), in large part due to the extraordinary radiation of insects. Remarkably, however, we have only a rough idea how many species exist either on land or in the oceans, possibly comprising some 8 to 9 million eukaryotes in total (Mora et al. 2011), with most of them still awaiting formal description. The challenge of identifying and explaining the seemingly endless variety of life on Earth remains one of the most profound tasks in biology, and perhaps in the sciences as a whole.

Apart from the sheer magnitude of biodiversity, its prominent spatial patterns have long been of interest. The nonrandom spatial distribution of biodiversity is apparent even to the casual observer. It is likely that early hominids who traveled and expanded their reach across different biomes would have already been cognizant of the substantial differences between the richness of ecological communities at large scales and their broad relationship to latitude, altitude, and moisture regimes. Ever since naturalists began describing, organizing, and documenting the variety of species in more detail, some very general patterns have emerged. For many taxonomic groups surveyed on land, including plants, vertebrates, and insects, species richness (the number of species found in a particular area at a given time) peaks in the wet, warm tropics and falls off sharply toward higher latitudes and altitudes (Gaston 2000). This *first-order pattern* of global biodiversity (Krug et al. 2009) is exceedingly well documented (Hillebrand 2004), but poorly understood from a theoretical perspective, with a raft of possible mechanisms proposed (Rohde 1992). In the oceans, similar gradients in species richness have been observed from the tropics to the poles in some well-studied coastal taxa (Stehli et al. 1967; Roberts et al. 2002). Open-ocean (*pelagic*) species, from zooplankton to whales, have been analyzed more recently and often display a different pattern that peaks at subtropical or even temperate latitudes (Tittensor et al. 2010). Deep-sea taxa, while sparsely sampled, appear different again (Woolley et al. 2016). Explaining these large-scale patterns has become a core question in ecology and evolutionary biology, often addressed through correlative methods linking hypothesized mechanisms to environmental variables. Although such approaches can help to identify possible drivers, they cannot necessarily distinguish between them or shed light on the mechanisms involved. An alternative route for enhancing our understanding of the spatial distribution of global biodiversity is through developing a model that enables, or at least proposes, a mechanistic understanding of the fundamental processes structuring global patterns in biodiversity, an ideal that has been called the "holy grail of modern biogeography and macroecology" (Gotelli et al. 2009).

Clearly, fundamental questions about species diversity are not just interesting to ecologists but are also central to our understanding of the world we live in and how we relate to it. Species diversity is akin to a periodic table of biology: it provides the fundamental building blocks for the ecosystems we all inhabit. How is the global richness of species organized, and how does it vary across taxa and through space and time? What are the environmental factors that may drive this variation and to what extent are these factors influenced by human perturbations? And finally, can we explain this bewildering variety from simple ecological theory and provide a more mechanistic understanding of the fundamental distributional patterns of life? These questions have long been at the core of ecology, and they form the focus of this book.

1.1. INTEGRATING LAND AND SEA

A key premise of our work here is that a more comprehensive understanding of global biodiversity can be gained only by overcoming a disciplinary divide that has

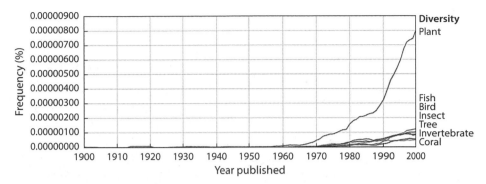

FIGURE 1.1. Mentioning of species diversity in published books. Plant diversity is referenced for around three times longer than and at least eight times as often as any other group. Note the increasing mention of other species groups only after around 1970, when spatial biodiversity science further expanded to other realms. Data from Google ngram project (https://books.google .com/ngrams).

long separated the study of biodiversity on land and in the sea. Since humans are land animals, we have often been primarily concerned with the terrestrial organisms that we can more easily observe. This is also reflected in the history of ecological study. Many scientists who shaped the foundations of our field, such as von Humboldt, Darwin, Wallace, Elton, Hutchinson, and MacArthur, had a terrestrial or sometimes freshwater focus, but they rarely considered the oceans in detail (Darwin's beautiful monographs on corals and barnacles being among the notable exceptions). These pioneers also shared a keen interest in the forces that shape patterns of species richness. Hence, published work on this topic is heavily biased toward the land. A quick online search confirms that there are about 30% more scientific articles and books devoted to the biodiversity of a single terrestrial habitat type—forests—alone than to marine biodiversity. Likewise, the mention of "plant diversity" in all published books, scientific or otherwise, goes back at least to the beginning of the twentieth century, whereas marine species groups appeared much later in this context and have only slowly risen to more prominence (fig. 1.1).

In focusing on the oceans in addition to the land in this book, we treat different marine habitats as further "replicates" of global biodiversity patterns: different from terrestrial, but potentially organized (at least at some fundamental level) by similar principles (fig. 1.2). At the very least, this assertion is a null-hypothesis to be confronted with data. At the very best, it increases the "degrees of freedom" when we test our ability to understand and predict the first-order patterns of biodiversity on our planet. As the marine environment is less familiar to most, it tends to challenge our assumptions about how the world works, and allows us to critically examine ecological concepts that have been developed with a largely terrestrial focus. Are these patterns, ideas, hypotheses, and theories truly general

FIGURE 1.2. Biodiversity across marine and terrestrial habitats provides four largely independent "replicates" with which to explore patterns of biodiversity. Examples of (A) coastal; (B) pelagic; (C) deep sea; and (D) terrestrial diversity are shown. Credits: (A) Ami Gur; (B) Samuel Blum; (C) NOAA; (D) Frank Brodrecht.

to all systems? It is our assertion that ecology, and its theoretical foundations, should be general, no matter whether its subjects happen to be wet or dry.

At the same time, our—and indeed society's—current interest in biodiversity is much more than academic. The rapid erosion of biodiversity—both on land and in the ocean—is cause for much concern, as it threatens individual species with extinction (Dirzo et al. 2014; McCauley et al. 2015), and ecosystems with loss of functionality, essential services, and resilience (Worm et al. 2006; Hooper et al. 2012). There is growing awareness of this biodiversity crisis as a defining problem of our time (Tickell 1997), with international policies being shaped in an attempt to slow and ultimately reverse the rate of loss (Tittensor et al. 2014). This has also resulted in renewed interest in the fundamental processes that give rise to, maintain, or threaten, biodiversity at local, regional, and global scales.

Some of these fundamental processes can be further unveiled by contrasting and comparing current patterns of and changes in diversity to those that have unfolded through deep time (Jablonski et al. 2006; Valentine and Jablonski 2015; Yasuhara et al. 2015). By carefully dissecting the paleontological record, researchers have been able to trace patterns of species richness throughout Earth's history, learning how they have changed as the environment around them became altered by geological, biological, and climatic forces (Renema et al. 2008; Krug et al. 2009; Yasuhara et al. 2012; Jablonski et al. 2013; Mannion et al. 2014). At present, human beings are exerting pressures and inducing changes as pronounced as many of these deep-time processes, and are certainly a dominant force shaping life on Earth (Waters et al. 2016). As an example, CO_2 is now emitted at a rate ~20 times greater than the fastest CO_2 emission rates in recorded geological history (Zachos et al. 2008), and is warming the planet at record speed. Yet, our ability to predict the future of biodiversity is both more limited and more uncertain than our understanding of its past (Sala et al. 2000). We posit that a more general understanding of global biodiversity patterns, and in particular the ecological and evolutionary forces that shape them, may help us to forecast biological reconfigurations in both the short and the long term under a given rate and pattern of environmental change.

1.2. A BRIEF HISTORY OF BIODIVERSITY RESEARCH

It is often instructive to trace the origin and evolution of a question or an idea through history. Observations on patterns of species richness at local scales likely date back to the beginnings of human inquiry. Traditional hunter-gatherer societies, for example, acquired and still hold detailed knowledge of many species and their distributions (Cotton 1996), and devised observation-based heuristics to explain such patterns. Early scientific inquiry was exemplified by Aristotle's strikingly detailed observations on both terrestrial and marine animals, particularly their morphology and distribution. These important observations were not synthesized in much greater detail until the Age of the Enlightenment. Specifically, Linnaeus's *Systema Naturae* (Linnæus 1758) established taxonomy and systematics, and enabled a more structured inquiry into patterns of species richness and how they unfolded based on the relatedness of individual organisms. Subsequently, major scientific expeditions were organized, discovering new species and ecological communities as they surveyed land and ocean areas around the globe. For example, Alexander von Humboldt and Aimé Bonpland's expeditions to South America of 1799–1804 initiated modern biogeography, while the *Challenger* expedition of 1872–1876 laid the foundations of oceanography, and catalogued over 4000 unknown species from marine waters around the world. A common thread running through this "era of discovery" was that it became increasingly evident that our planet held far more forms of life than had hitherto been anticipated, and that much of it was concentrated in tropical forests on land, or contained in structured habitats such as kelp forests and coral reefs in the ocean. In the marine realm, however, the pace of discovery was generally slower, particularly in the vast reaches of the open ocean and the deep sea, the latter long presumed to be devoid of life.

Darwin's theory of evolution (Darwin 1859) provided another milestone. The theory was elegant in that it required only two fundamental processes—specifically, the generation of variation in species' traits and the forces of natural selection acting upon that variation—to explain the emergence of new species, and by extension all of biodiversity. Although the *Modern Synthesis* suggests that other processes, such as drift, are also important, the ramifications of Darwin's theory as a structural foundation for anchoring an understanding of the spatial distribution of biodiversity are obvious. Yet, there was little theory available or developed at the time to understand the processes shaping the distribution of those species across regions and around the globe.

Modern biodiversity science was arguably born out of Gene Evelyn Hutchinson's "Homage to Santa Rosalia" (Hutchinson 1959), in which he posed the question of how the great variety of observed species could coexist in a given environment while competing for a few limiting resources. Examples include the astounding richness of plankton species that can coexist in a single drop of water, or the large number of tree species found in a patch of tropical forest. Hutchinson's student Howard Sanders took this question to the deep sea, where he and Robert Hessler discovered surprising levels of macrofaunal diversity that were at the time perceived to perhaps match those of hyperdiverse tropical forests and coral reefs (Hessler and Sanders 1967). Soon Sanders, among others, was formulating theory to explain these observations (Sanders 1968). The *stability-time hypothesis*, which was borne out of these observations, was one of the first theories specifically developed to explain observed biodiversity patterns; it related large-scale differences in species richness to the severity and frequency of disturbances. This work partially inspired several decades' worth of research into marine biodiversity patterns, both from biologists, focusing on coral reef and deep-sea macrofauna (McCoy and Heck 1975; Grassle and Maciolek 1992; Rex et al. 1993; Roberts et al. 2002; Brandt et al. 2007), and geologists, focusing on bivalves, foraminifera, and other microfossils (Ruddiman 1969; Stehli et al. 1969; Rutherford et al. 1999; Valentine and Jablonski 2015).

In a parallel development, MacArthur and Wilson published their *Theory of Island Biogeography*, arguably the first mathematical theory of biodiversity patterns (MacArthur and Wilson 1967). While originally focusing on the immigration and extinction of terrestrial species on oceanic islands, it was subsequently applied to other isolated habitats such as lakes, mountaintops, and forest fragments. A later extension of this theory, *The Unified Neutral Theory of Biodiversity and Biogeography* (Hubbell 2001), which built on MacArthur and Wilson's work, was heavily informed by observations in tropical forests. When its predictions were tested in other habitats both marine (Dornelas et al. 2006) and terrestrial (McGill et al. 2006), however, results were mixed, and its generality remained unclear. The *Metabolic Theory of Ecology* is another, more recent theory that makes general predictions about changes in community structure and consequently biodiversity in relation to temperature (Brown et al. 2004). On its own, it also falls short of capturing a majority of diversity gradients on land (Hawkins et al. 2007), but it has not been evaluated in comprehensive detail for marine environments.

The new millennium brought about an era of large-scale data integration and synthesis, spurred by the growth of macroecology as a prominent subfield in ecology (Brown 1995; Gaston 2000). The First Global Census of Marine Life (2000–2010), for example, built on Fred Grassle's and Jesse Ausubel's bold vision to systematically chart and understand marine biodiversity patterns across all the many different habitats, from shallow reefs to abyssal plains (Ausubel 1999; Grassle and Stocks 1999). Grassle was a student of Sanders, and thus it is conceivable that the vision of a systematic and global marine census may ultimately be traced back through a direct academic lineage to Hutchinson's seminal thoughts on patterns of species diversity. The Census of Marine Life was by far the largest initiative of its kind and certainly invigorated interest in marine biodiversity, both in academia and in the public eye through regular media coverage of its many spectacular and photogenic discoveries. Scientifically, it fostered an interdisciplinary, comparative, and highly collaborative approach that culminated in a series of synthetic papers on the large-scale patterns of species distribution, abundance, and richness across different marine habitats (Tittensor et al. 2010; Block et al. 2011; Mora et al. 2011; Ramirez-Llodra et al. 2011). At the same time, parallel synthetic developments in terrestrial ecology led to similar efforts at empirical synthesis, focusing largely on patterns in plant (Kier et al. 2005; Kreft and Jetz 2007) and vertebrate species richness (Jetz and Rahbek 2002; Grenyer et al. 2006; Jetz and Fine 2012). Finally, the creation of dedicated institutions such as the National Center of Ecological Analysis of Synthesis (NCEAS) spurred the search for general ecological principles to be extracted from these newly documented global patterns. Most recently, online databases such as the Encyclopedia of Life (http://eol.org/); the Map of Life (https://www.mol.org/); the International Union for Conservation of Nature (IUCN) Red List (http://www.iucnredlist.org/); the Global Biodiversity Information Facility (http://www.gbif.org/); and the Ocean Biogeographic Information System (http://www.iobis.org/) have enabled even better integration of biological and environmental data, facilitating further syntheses of observed patterns of species richness and community structure.

Despite these exciting advances, however, biodiversity research has not yet produced a testable body of theory that can coherently explain the manifold patterns of global biodiversity on land and in the sea. Yet, the time seems ripe to attempt such a synthesis. On the empirical side, field ecologists have compiled an unprecedented number of observations across multiple taxonomic groups along with associated environmental predictors that might be related to underlying processes. There also exists an extensive list of hypothesized drivers and mechanisms (Rohde 1992), but no cohesive model to reconstruct observed patterns of biodiversity from such mechanisms at the global scale. Especially challenging is the fact that some of the hypothesized drivers, such as temperature, can operate through both ecological (for example, niche constraints) and evolutionary (for example, speciation rates) mechanisms, and correlative models provide no way to separate these effects. Another obstacle is that much relevant theory (for example, island biogeography theory, metabolic theory, or indeed neutral theory) has been developed with a terrestrial focus, and its applicability to marine habitats has been less extensively explored in terms of spatial distributions of biodiversity. Integrating empirical observations with mechanistic models both on land and in the ocean may provide a clearer picture of both their generality and their limitations.

1.3. GOALS AND STRUCTURE OF THIS BOOK

Our goal in this book is to construct an integrated understanding of large-scale patterns of marine and terrestrial biodiversity at regional to global scales (hundreds to tens of thousands of km). Although we recognize that the term *biodiversity* is used to encompass all levels of biological variation (UN 1992), here we shall use it primarily to describe the number of species in a given community (species richness); other aspects of biodiversity such as phylogenetic or functional diversity will remain an area for future research. We begin by describing and synthesizing the state of scientific knowledge of species richness patterns for coastal, pelagic (open ocean), and deep-sea (>2000 m depth) environments, and compare these to patterns found on land and in freshwater (chapter 2), treating these major habitats as largely independent "replicates" across which we strive to understand contrasting patterns of diversity. We then discuss and analyze the environmental correlates that best explain these patterns (chapter 3). Based on this empirical understanding, we construct a simple theoretical model that attempts to explain the present latitudinal distribution of species, in an idealized ocean (chapter 4). In starting with a "minimal realistic model" and then building on it to produce something more closely aligned with observed patterns, we find that we need to unify three preeminent ecological theories—namely, neutral, niche, and metabolic theory. From this emerge a number of predictions on species richness, but also range size, niche breadth, and other macroecological patterns that are explored in some detail. When confronting our model with empirical data (chapter 5), we derive surprisingly realistic predictions of species richness for known marine and terrestrial species groups on a global grid that includes both the continents and the oceans. On the more applied side (chapter 6), we use our empirical synthesis to map out possible priority areas for conserving biodiversity at large scales, and employ our theoretical model to explore scenarios of long-term changes in future biodiversity patterns in response to projected global warming. In the long term, it is our hope that our thoughts will in some way contribute to a broader, unifying synthesis in biodiversity science (chapter 7).

The ideas presented in this book are necessarily circumscribed by inherent biases in our collective knowledge of different species groups. We tend to be familiar with larger-bodied taxa that are easily observable, or that we find beautiful, nutritious, or fearsome. Much less information is available, particularly when seeking global coverage, for those taxa that are smaller, less visually interesting, or less obviously useful for humankind. Microbial organisms, for example, are critically important in regulating biogeochemical processes at the planetary scale (Falkowski 2012), yet our knowledge of their biodiversity remains very limited (Sunagawa et al. 2015). Less well researched taxa may display different spatial patterns of biodiversity from those with which we are more familiar. We are unable, unfortunately, to overcome this limitation, and can only work with the empirical information that we have available at this point in time; future discoveries and empirical compilations can be used to test, revise, or overturn the ideas in this book. In addition, our work maintains a consistent focus on species richness, and treats species evenness, functional diversity, species turnover, and community composition only in passing. This is most certainly not for a lack of interest but due to the inherent limitations involved in covering a broad range of taxa and patterns in a single book.

We also focus deliberately on large scales, for the pragmatic reason that our knowledge at large spatial grains (hundreds to thousands of km) tends to be more complete, on average, than at finer grains (Mora et al. 2008). The idea is to capture the first-order latitudinal and longitudinal patterns of species richness, its peaks and troughs, rather than its detailed regional variation, which might be structured by different driving forces. We realize that this approach may disappoint some, who would like to see a more detailed representation of localized marine and terrestrial biodiversity patterns, as well as a detailed explanation of smaller scale variation, but this would necessitate a very different focus for this volume. We fully acknowledge that we present only generalities and that potentially important exceptions and scale-related differences exist, and we direct the reader toward more regional studies wherever possible.

CHAPTER TWO

Observed Patterns of Global Biodiversity

As a first step in our exploration of global biodiversity patterns, we attempt a synthetic review of observed patterns of species richness across four major environmental realms (coastal, pelagic, deep sea, and land), while their ecological drivers and environmental correlates are discussed subsequently in chapter 3. Our synthesis is aided by an unprecedented growth in spatial ecology over the last two decades, leading to new insights into the geographical patterns of biodiversity. This growth has been spurred in part by new tools for collecting, storing, and analyzing data (for example, remote sensing Geographic Information Systems [GIS]), and voluminous new biodiversity databases (for example, the Ocean Biogeographic Information System [OBIS] and the Global Biodiversity Information Facility [GBIF]), but also by conceptual advances in macroecology (Brown 1995) and biogeography (Beaugrand 2014). This growth in spatial ecology parallels earlier, but largely unconnected, developments in geology (Ruddiman 1969; Stehli et al. 1969; Stehli and Wells 1971) and paleobiology (Balsam and Flessa 1978)—fields that are equally concerned with large-scale biogeography, albeit from a different perspective. Despite these advances, there has been little exchange or synthesis between terrestrial and marine studies of global biodiversity, and across paleontological and recent data, with the relevant literature remaining somewhat separated by disciplinary boundaries (but see Stehli et al. 1969; Rohde 1992; Gaston 2000; Webb 2012).

Our aim in this chapter is to summarize and synthesize known biodiversity patterns and analyze them for congruency over space and time. To do so, we focus on studies with patterns derived directly from empirical data, rather than those that spatially extrapolate based on environmental relationships or assumed niches. We treat such modeling studies only in passing. Our rationale is that spatially modeled patterns are generally based either on niche models with assumed environmental drivers or on spatial prediction where gaps in sample data are filled and interpolated based on environmental determinants. This is relevant because one key goal of this book is to develop theory and models to better understand the underlying drivers of global biodiversity patterns from first principles. Assessments of these relationships can be performed only at locations with empirical data; to do otherwise would lead to circularity in reasoning (environment is used to predict richness, which is then correlated with environmental determinants). We do, however, sometimes include maps that have been spatially extrapolated, some of which use environmental covariates. This is for visualization purposes only: any analysis we conduct in this or the following chapters is based exclusively on the raw observational data with spatially extrapolated locations excluded.

Generally, we limit our discussion to macroecological patterns at continental to global scales (thousands of km), as patterns and drivers of species richness on smaller scales are often idiosyncratic and likely more reflective of local processes (Belmaker and Jetz 2011). Scale-dependence in the relationship between species richness and the environment has been identified multiple times (Storch et al. 2007), and the factors that control local community structure and biodiversity are treated in some detail elsewhere (Tilman 1982; Hubbell 2001; Storch et al. 2007). In this volume, we have consistently standardized the available data on an equal-area 880 × 880 km grid, which was found to capture global patterns well and reach a good compromise between data availability and information content (Tittensor et al 2010), as well as reflecting the true state of knowledge in a conservative manner (Mora et al 2008). It also reflects our intent to model the processes driving large-scale (as opposed to fine-scale) patterns of biodiversity. While species richness information for some taxa, particularly on land, is available at finer resolutions, we chose to make all patterns comparable and consistently scaled. In addition to spatial patterns on the equal-area grid, we also report the latitudinal gradient in species richness for each species group, which reflects average richness across all nonzero cells in each latitudinal band. The interested reader can follow up on the relevant citations to view patterns at the original resolution. We generally do not imply that our findings can always be replicated at finer scales, and we will discuss questions about processes across scale where appropriate.

Given the questions that we ask in this volume, we are unavoidably bound to focus on taxonomic groups for which well-sampled global diversity patterns are available. Typically, these are recognizable species of low to moderate richness that often fossilize well, such as foraminifera, bivalves, or vertebrates. Spatial richness information for these groups may be reported at various levels of taxonomic resolution, from families to phyla. We generally retain the original taxonomic level, but present a sensitivity analysis in section 2.6, where we examine the effect of taxonomic resolution on documented patterns of species richness. We further recognize that a large fraction of the planet's biodiversity remains beyond our analytic reach, particularly most invertebrates and microorganisms, though the advent of high-throughput DNA methods is beginning to alleviate this (with the caveat that the species concept breaks down at the level of bacteria and viruses).

Throughout this chapter, we also retain our focus on the simplest measure of biodiversity—namely, species richness. Other aspects such as species turnover, functional diversity, evenness, and genetic diversity are acknowledged as clearly important, but are not sufficiently well known at the scales that we examine to enable a major synthesis across taxonomic groups on land and in the sea. We separate coastal, pelagic, deep-sea, and terrestrial richness patterns, since these major realms are sufficiently discrete as to necessitate independent treatment (Gaston 2000; Tittensor et al. 2010; Woolley et al. 2016), and can provide informative contrasts, shedding insight on generalities and differences. Within these broad habitats, we separate by major taxonomic groups, discussing and comparing patterns for plants, invertebrates, and vertebrates.

2.1. MARINE COASTAL BIODIVERSITY

Coastal habitats, here defined as the shallow waters stretching from the intertidal zone to the edge of the continental shelf (0–200 m depth), represent the most accessible marine environments for humans and hence the best known. As a transition zone between land and sea, they are strongly influenced by a combination of terrestrial and marine processes. Land runoff and rivers increase the availability of nutrients, and on average result in a much higher (10- to 100-fold) net primary productivity (NPP) compared with most open-ocean environments. Coastal NPP frequently ranges in the order of 500 to 2500 g C m^{-2} a^{-1}, which is similar to productive grasslands and forests (Lieth and Whittaker 2012). Also similar to the land, we find high habitat diversity and structure here, often created or enhanced by foundation species such as corals, macroalgae, seagrasses, or mangroves. Many coastal environments show sharp physical gradients both vertically and horizontally in temperature, salinity, productivity, and depth that enhance the heterogeneity provided by biogenic habitats. These sharp environmental gradients and complex habitats can affect both ecological and evolutionary processes that in turn structure biodiversity, resulting in complex local gradients and heterogeneity. Aside from a dynamic physical environment, coastal habitats also experience significant impacts from a growing human population that has chosen to settle preferentially near the world's seashores. The effects of exploitation, pollution, and habitat conversion are all concentrated in coastal environments, particularly in the northern hemisphere (Halpern et al. 2008), and over time have profoundly altered patterns of biological diversity, habitat structure, and productivity (Lotze et al. 2006). In summary, the most outstanding features of coastal environments when compared to the pelagic ocean and the deep sea, and akin to those on land, are complex habitat structure, high nutrient input and productivity, and growing human impacts, all of which influence the distribution of coastal biodiversity.

2.1.1. Coastal Invertebrates

Large-scale patterns of coastal biodiversity were first described for bivalves (Mollusca; Class: Bivalvia), initially at a regional scale (Thorson 1952, 1957), and then globally in the foundational studies by Stehli and Sanders (Stehli et al. 1967; Sanders 1968). Bivalves are particularly useful specimens, as they are found in all coastal habitats, are easily sampled, and are reasonably taxonomically tractable (~10,000 known extant species). Moreover, this group covers much of the geological history of metazoans, having emerged during the Cambrian explosion of animal diversity about 540 Ma ago. Bivalves fossilize well, which sheds light on their evolutionary history, and helps to interpret the emergence of biogeographical patterns over time. It is worthwhile to read these classic studies closely, as they reveal luminous insights into some of the first-order patterns, and likely drivers, of marine biodiversity. By fitting diversity contours around samples of bivalve diversity from just 36 locations (fig. 2.1A), Stehli and coauthors first identified two prominent centers of diversity (later called *hotspots*): one in the tropical Indo-Pacific centered on the Indonesian archipelago, and a secondary one in the tropical Eastern Pacific off the northern coast of Central America (Stehli et al. 1967). These species-rich features were almost perfectly centered on the equator, and were also seen at the genus and family level. Other primary features of the overall patterns were (1) steep latitudinal gradients with maximum diversity in the tropics, (2) less steep longitudinal gradients away from the Indonesian and Central American hotspots, and (3) low diversity in the Arctic and Southern Oceans (fig. 2.1).

Progressive updates and the inclusion of new data have largely confirmed these earlier results (Crame 2000; Valentine and Jablonski 2015), with the exception that richness appears to peak slightly north of the equator, but still in tropical waters (fig. 2.1B). These results prefigure the canonical pattern of biodiversity later identified for a variety of other coastal marine taxa (Tittensor et al. 2010). This consistency may hint at some generalities in spatial biodiversity patterns that can be readily gleaned from incomplete and limited data at a coarse spatial grain (see fig. 2.1). Another interesting observation from this early work relates to the fact that steeper latitudinal gradients were seen in more recently evolved bivalve taxa; this may suggest a tropical origination for most bivalves and a long dispersal period (10^6 to 10^8 years) by which newly formed species groups, or *clades*, reached the poles via adaptive radiation (Crame 2000; Jablonski et al. 2006; Jablonski et al. 2013).

A

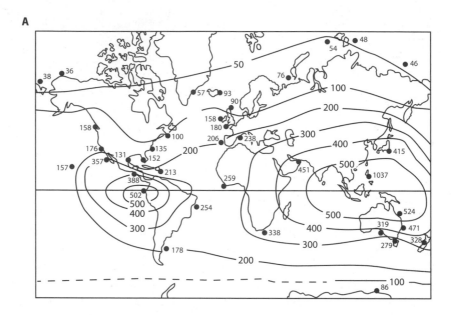

B

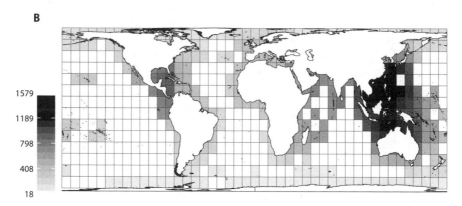

FIGURE 2.1. Diversity of recent bivalves. The first global map of marine biodiversity was published in 1967 for bivalves ((A) sampled richness of marine bivalves, after data in Stehli et al. 1967). Based on a limited number of samples from 36 locales, Stehli and colleagues synthesized a first-order global diversity pattern in coastal and continental shelf ecosystems. Remarkably, the general patterns have not changed significantly, although data availability has improved considerably since then ((B) total richness of marine bivalves, after data in Valentine and Jablonski 2015). Likewise, the pattern appears consistent when changing taxonomic resolution; near identical gradients were seen at the genus and family level (Stehli et al. 1967). Finally, it has emerged that very similar patterns are seen across most known coastal species groups, implying marked generality within and across taxa (figs. 2.2–2.4).

Apart from their groundbreaking work on bivalve species richness, Stehli and colleagues also analyzed the richness of reef-building corals (Stehli and Wells 1971), which form the most species-rich habitats in the coastal realm. Coral reefs are biogenic coastal habitats created by stony corals (Cnidaria; Order Scleractinia) that first evolved in the Triassic, about 240 Ma ago. Coral reefs are often likened to rainforests due to their astounding local diversity, although their spatial extent is much smaller, at about 260,000 to 600,000 km^2 worldwide, approximately 5% of total rainforest area, and only 0.2% of ocean surface area. Yet these reefs have among the highest diversity per unit area of any known habitat, much of it remaining undescribed (Bouchet et al. 2002; Bouchet 2006). As in rainforests, the largest number of species is not found among the structural species themselves (reef-building corals contain ~800 known species) but in the diverse fauna that inhabits these complex habitats. Birds and insects in a rainforest find an equivalent in the fishes and invertebrates on a coral reef.

In a similar manner to the bivalves, the diversity pattern for reef-building corals is centered on a global biodiversity hotspot in the tropical Indo-Pacific (fig. 2.2). Species richness declines with increasing distance from this center, the so-called coral triangle between the Philippines, Indonesia, and Australia that supports most known species of reef-building corals. A secondary hotspot in the Caribbean harbors only ~50 or so species of hard corals. Even in the earliest works on the topic, it was already noted that the coral triangle and Caribbean hotspots appear to have emerged independently of one another, as faunal overlap is very small (Stehli and Wells 1971). Moreover, the authors remarked that both hotspots were associated with the warmest sea surface temperature (SST) anomalies in the Indo-Pacific and Atlantic, respectively; in both instances, these occur in the western part of the basin, and somewhat north of the equator (Stehli and Wells 1971). The authors argued that water temperature may play a leading role in enabling high richness, as opposed to solar radiation, which peaks at the equator. The role of habitat area was also discussed, as both regions harbor large archipelagos with long coastlines, and complex spatial structure, but the relative importance of this was not resolved analytically. Furthermore, the authors noted that the age of surveyed genera increased with increasing distance from the centers, and young genera were predominantly found in both the Caribbean and western Pacific centers of diversity. This supported the hypothesis that these centers of high diversity were also centers of evolutionary innovation. Interestingly, the mean age of coral genera increases much faster with latitude than with longitude (Stehli and Wells 1971), implying slower evolutionary speeds at higher latitudes (and at lower temperatures), and maybe a slower spread of newly evolved genera across latitude when compared to longitudinal expansion. We will return to these environmental drivers in chapter 3.

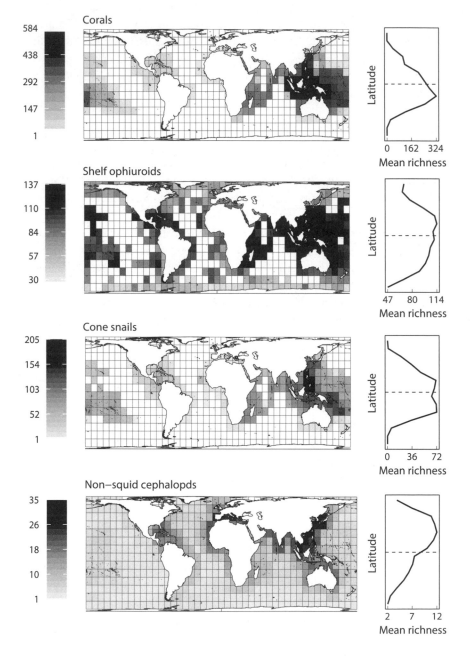

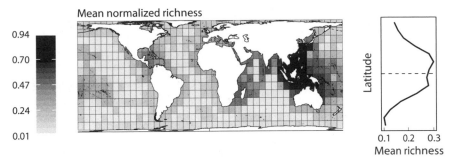

FIGURE 2.2. Coastal invertebrates. Shown are global species richness patterns and corresponding latitudinal gradients for corals, ophiuroids (brittle stars), cone snails, and coastal cephalopods in coastal and continental shelf waters <200 m depth. Data from Tittensor et al. (2010); IUCN (2016); Woolley et al. (2016). Note that the cephalopod pattern includes only commercial species. The bottom panel displays mean normalized richness, in which richness values for each taxon were scaled from zero to one and then averaged to standardize patterns irrespective of differences in total richness between taxa; normalized richness also includes the bivalves data from fig. 2.1B.

As corals provide habitat for a large number of associated mobile organisms, such as fish and invertebrates, it is interesting to investigate whether their diversity patterns match up with those of corals. Expert-derived maps of coral, reef fish, lobster, and gastropod diversity were compiled by Roberts et al. (2002). They all showed very similar spatial patterns (pairwise Spearman's rank correlations ranged from 0.78 to 0.89), again with a clear center of diversity in the coral triangle. Areas in the southern Philippines and central Indonesia were in the top 10% richest locations for all four taxa, and diversity decreased with increasing distance from this region. The secondary hotspot in the Caribbean was also visible across all taxa, although not quite as pronounced as in the corals. Did these diversity patterns of reef-associated fauna arise independently, through similar historical and environmental forcing? Or, alternatively, do corals act as foundation species that promote further diversification in associated groups independently of the environmental regime? Of course, these two explanations are not mutually exclusive. In a recent paper, it was hypothesized that coevolutionary relationships in which "diversity begets diversity" may potentially be an important factor in amplifying the effects of primary environmental drivers, such as temperature, on diversity (Brown 2014). We concur that while such coevolutionary relationships may enhance existing biodiversity gradients, they cannot generate them in the first place (Rohde 1992).

These general patterns observed for bivalves and corals hold surprisingly well across a number of other shallow-water invertebrate taxa, with few exceptions (Fischer 1960; Roy and Witman 2009; Tittensor et al. 2010), and suggest the presence of some very general structuring process or processes for shallow-water coastal diversity. Brittle stars (Echinodermata; Class: Ophiuroidea), cone snails (Mollusca; Family: Conidae), and coastal cephalopods (Mollusca; Class: Cephalopoda), for example, belong to different phyla with divergent evolutionary histories, ecological roles, and capacities for dispersal, yet their global diversity patterns peak around the same area in the tropical Indo-Pacific as corals and bivalves (though with some latitudinal variation in the peak, and sometimes a small decrease at the equator), and show similar overall gradients (see fig. 2.2). Of course, significant residual variability does exist, possibly due to those different evolutionary histories—for example, cone snails are relatively less diverse in the Atlantic, and cephalopods appear to show lower richness in the southern hemisphere, yet the general features of the coastal biodiversity pattern seem conserved across all marine invertebrate groups examined (see fig. 2.2). Somewhat surprisingly, this generality also extends to coastal marine plants (fig. 2.3) and vertebrates (see fig. 2.4, later).

2.1.2. Coastal Plants

Seagrasses (Plantae; Families: Posidoniaceae, Zosteraceae, Hydrocharitaceae, Cymodoceaceae) and mangroves (Plantae; Families: Acanthaceae, Combretaceae, Arecaceae, Rhizophoraceae, Lythraceae) are two unusual groups of flowering plants (angiosperms) that have adapted to life in the coastal oceans. They can colonize soft sediments, while corals settle only on hard substrates, and hence these marine plants represent important habitat-forming foundation species on surfaces where corals cannot grow. While mangroves have a similar tropical and subtropical distribution to shallow-water corals, seagrasses are also present in temperate environments. As noted 40 years ago (McCoy and Heck 1975), these three groups, although taxonomically unrelated to each other, show surprising similarity in their geographical diversity patterns, with clear peaks in the western tropical Pacific for mangroves and a similar, but more subtropical, pattern for seagrasses (see fig. 2.3). Similarly to corals (and bivalves), a secondary diversity hotspot is found in the Caribbean for both groups, and the correlation between these global patterns is high (Tittensor et al. 2010). The observed strong overlap in species richness has been hypothesized to arise from an ecological, rather than phylogenetic, relationship between them (McCoy and Heck 1975). It is thought that corals initially colonize hard surfaces and provide habitat for sediment-trapping filamentous and calcareous algae. The accumulating sediment is colonized by seagrasses, which further enhance sedimentation and build up substrate that can then be colonized by mangroves (McCoy and Heck 1975). Indeed, in many coral atolls, these groups are found in close proximity, and mangroves, seagrasses, and

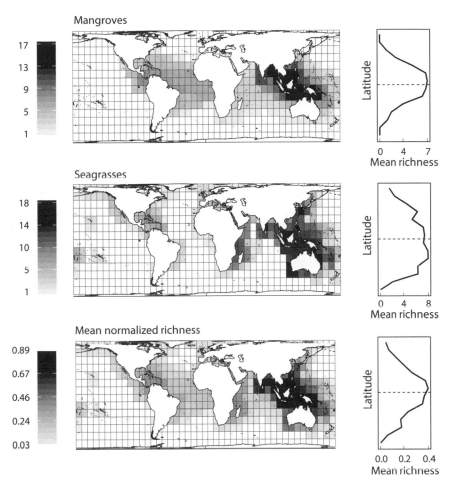

FIGURE 2.3. Coastal plants. Shown are the global species richness patterns and corresponding latitudinal gradients for the two groups of flowering plants that colonize shallow marine waters—namely, seagrasses and mangroves. Mean normalized richness pattern averaged across these groups is shown at the bottom. After data from Tittensor et al. (2010).

corals are sometimes used consecutively by the same reef fish species as larval, juvenile, and adult habitat, respectively (Mumby et al. 2004). Yet this potential ecological relationship and successional sequence does not explain why spatial patterns of species richness (as opposed to presence or biomass) are so strongly overlapping, and again "diversity begets diversity" cannot be the whole explanation. Possibly these species groups may relate in a similar way to the environmental factors that shape the habitat in which they co-occur. These include the extent and complexity of coastal habitats found in the Indonesian-Australian and Caribbean archipelagos and the fact that these regions experience the highest average sea surface temperatures in their respective ocean basins. These factors will be analyzed and discussed in detail in chapter 3.

In temperate coastal environments, most shallow corals and mangroves cannot survive due to their limited tolerance of cold water (McCoy and Heck 1975; Hutchings and Saenger 1987), though see Cairns (2007) for a discussion of cold-water corals. Instead, macroalgae (on rocky shores) and seagrasses (on soft sediments) provide shallow temperate biogenic habitats. Reflecting this pattern of distribution, seagrass species richness is skewed more poleward when compared to mangroves (see fig. 2.3) or corals (see fig. 2.2). In contrast to seagrasses and mangroves, which are less speciose, we do not presently have a complete picture of macroalgae species richness patterns on a global scale; this is partly due to higher richness, problematic species identification, and fluid taxonomy at the species level. Genus richness, however, has been assessed, and peaks at progressively higher latitudes for the three main clades: green algae tend to center in the tropics (Archaeplastida; Division: Chlorophyta), red algae (Archaeplastida; Division: Rhodophyta) at intermediate latitudes, and brown algae (Ochrophyta; Class: Phaeophyceae) at temperate latitudes, such as off South Australia, Europe, and Japan (Kerswell 2006; Keith et al. 2014). At the species level, the Bryopsidales, a speciose order of mostly turf-forming green algae, has been analyzed in more detail; like other Chlorophyta, this group shows highest species diversity in the tropics, and specifically in the tropical Indo-Pacific. A secondary hotspot is found in the Caribbean, but with much lower diversity in the Atlantic overall (Kerswell 2006). As such, the distribution of species richness does closely follow that of corals and mangroves. The reason might be that Bryopsidales tend to be largely reef-associated, whereas many other macroalgae thrive on rocky surfaces not occupied by reef-forming corals, and hence tend to be distributed in a more poleward manner (Kerswell 2006; Keith et al. 2014). In summary, macroalgae seem to show a variety of richness patterns, ranging from tropical distributions similar to corals, to temperate distributions similar to seagrasses (see fig. 2.3).

2.1.3. Coastal Vertebrates

Expanding the scope from largely sessile invertebrates and plants to more mobile vertebrates (Chordata; Subphylum: Vertebrata) provokes the question of whether observed diversity patterns may blur or weaken due to increased capacity for migration and dispersal and the lack of hard boundaries in a continuous fluid marine habitat. Vertebrates have evolved as mobile consumers, with bony fishes representing the most speciose taxon, containing half of all known vertebrate species. The modern bony fishes (Class: Osteichthyes) probably evolved in the late Silurian, about 416 Ma ago. They inhabit surface to abyssal depths and open-ocean, coastal and estuarine, as well as freshwater habitats. About 16,000 to 17,000 species have been described in the oceans (Mora et al. 2008; Eschmeyer et al. 2010). According to an analysis of species discovery curves, this comprises an estimated 79% of total marine fish richness (Mora et al. 2008), suggesting that fish diversity is much more completely known than most other marine taxa (Mora et al. 2011).

Somewhat surprisingly, given their contrasting dispersal ability and evolutionary history, coastal bony fishes again yield a similar diversity pattern as corals, mangroves, and seagrasses, with a pronounced diversity peak in the tropical Indo-Pacific, although with small bimodal peaks north and south of the equator (fig. 2.4). However, latitudinal and longitudinal gradients of species richness appear less steep in fish, possibly relating to enhanced dispersal, and to the fact that coastal fishes cover a wider latitudinal range than either corals or mangroves. Secondary hotspots for coastal fish diversity occur in the western Indian Ocean, the Caribbean, and the eastern tropical Pacific (see fig. 2.4). These general patterns are seen across a large number of regional to global studies looking at fish diversity (see review by MacPherson et al. 2009). Note that the Eastern Tropical Pacific and Caribbean hotspots have a shared evolutionary history (prior to the closure of the Isthmus of Panama 3 Ma ago), and still show large overlaps in species composition; about 35% of Eastern Pacific fish genera are shared with, and only with, the Caribbean (Rosenblatt 1967).

There is an interesting contrast, however, between coastal bony fishes and coastal sharks (Class: Chondrichthyes; Superorder: Selachimorpha; see fig. 2.4), which evolved in the Silurian (440 Ma ago). The majority of the 508 or so known shark species are associated with coastal habitats at least for part of their life cycle (Lucifora et al. 2011), but these coastal sharks reach their highest diversity at somewhat higher latitudes than coastal teleost fishes: distinct peaks of shark species richness are found at around 20 degrees latitude North or South in the Western Pacific and Eastern Indian Ocean (see fig. 2.4). As such, their longitudinal diversity pattern is similar to other coastal species, but the latitudinal pattern is spread out farther toward the subtropics. This might be related either to the larger average size of sharks compared with bony fishes, which may translate into enhanced dispersal and larger ranges, or the longer evolutionary age, which allows more time for dispersal and adaptation to cooler climates (MacPherson et al. 2009). Indeed, several species of lamnid sharks, such as white (*Carcharodon carcharias*), porbeagle (*Lamna nasus*), and salmon sharks (*Lamna ditropis*), have evolved partial endothermy, which allows for a more poleward distribution, at least for part of the year (Block et al. 2011).

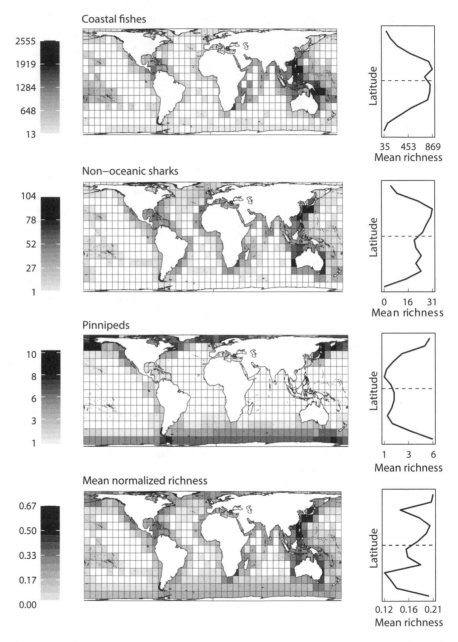

FIGURE 2.4. Coastal vertebrates. Shown are the global richness patterns and corresponding latitudinal gradients for coastal bony fish, sharks, and pinnipeds (seals, sea lions, and walrus). Mean normalized richness pattern averaged across these groups is shown at the bottom. After data from Tittensor et al. (2010).

Marine mammals (Class: Mammalia), of course, have perfected endothermy, and hence possess a large competitive advantage in cold-water environments by maintaining high metabolic rates and burst speeds. As such, it may not be a surprise to find that pinnipeds (Suborder: Pinnipedia—seals, walrus, and sea lions; 33 extant species) show a decidedly more poleward distribution in species richness (Tittensor et al. 2010). This recently evolved group of fin-footed marine mammals (origination ~30 Ma ago) has diversified mostly in temperate to polar habitats. Molecular evidence supports a monophyletic ancestry in the New World—specifically, the Northern Pacific Coast of North America (Arnason et al. 2006). As cold-adapted specialists, the pinnipeds have subsequently spread to Arctic and Antarctic waters, but are mostly absent from the tropics (except in Central and South America, where coastal upwelling cools surface waters, and in the endemic Mediterranean, Caribbean, and Hawaiian monk seals; Tribe Monachi). Peak species richness is observed in subarctic and Antarctic environments, where pinnipeds can be dominant predators (see fig. 2.4). As such, the pinnipeds are similar in their richness patterns to other cold-adapted taxa—for example, the Antarctic icefishes (Class: Osteichthyes; Suborder: Notothenioidei) and penguins (Class: Aves; Order: Sphenisciformes). These taxa likely originated in cold-water habitats, unlike most other groups, and then secondarily spread to warmer environments from there. Thus, while patterns for their parent taxa (marine mammals, fishes, and birds) all have tropical centers of diversity, these cold-adapted subgroups have a different diversity pattern likely due to their unique adaptations and a center of origin at higher latitudes.

2.1.4. Synthesis

In summary, the coastal realm is, in terms of marine environments, the closest to us spatially, and also offers the closest comparison to the land, as habitat structure and nutrient supply are more similar to terrestrial analogues than other marine environments. Empirically, we observe substantial overlap in the distribution of diversity among different species groups, including invertebrates, plants, and vertebrates. A majority of these groups show maximum diversity in the Indonesian-Australian (Indo-Pacific) and Caribbean (Atlantic) archipelagos, respectively. Latitudinal peaks in richness are generally tropical (though not necessarily equatorial) or subtropical, except for cold-water adapted subgroups. These generalizations are supported when averaging normalized diversity across groups (see figs. 2.2 to 2.4). For all groups examined here, total richness in the Atlantic is lower on average than in the Indo-Pacific, possibly reflecting the younger geological age of the Atlantic, which formed only 130 Ma ago. The large cluster of species richness in the central Indo-Pacific is striking and has no equivalent in the western Indian Ocean or eastern Pacific, for example. This pronounced longitudinal gradient is also markedly different from pelagic, deep-sea, and terrestrial taxa, which are discussed in the next sections.

2.2. MARINE PELAGIC BIODIVERSITY

The pelagic realm, here defined as the open waters that extend beyond the continental shelf, is a defining feature of our planet, yet few people ever get the chance to truly explore it. Pelagic waters encompass about two-thirds of the planet's surface and are characterized by great depth ranges (up to ~11,000 m). As such, this represents by far the largest volume of any habitat on Earth, yet our knowledge lags far behind what we know about biodiversity in coastal waters. In the pelagic ocean, most data pertain to the euphotic zone—that is, the top 200 m layer of the water column that receives sufficient sunlight for photosynthesis. Deeper pelagic waters, particularly those beyond 2000 m depth, are much less well studied, and are poorly understood in terms of their biodiversity. Even the largest known deep-water pelagics, such as giant squid (*Architeuthis* sp.) have only recently been documented for the first time in their native habitat (Roper and Shea 2013). For this reason, we largely focus on the euphotic (sunlit) pelagic zone here (top 200 m), with some discussion of mesopelagic taxa (200–2000 m). Deep-water taxa (>2000 m depth) will be discussed in the next section.

The pelagic environment experiences relatively little direct influence from the land and the sea floor, and consequently receives low input of macronutrients. The availability of limiting nutrients—particularly, nitrogen, phosphorus, and iron— is generally dependent on the physical structure of the water column and can be enhanced by local upwelling of nutrient-rich deep water—for example, in the equatorial upwelling regions—or around sharp frontal zones (Moore et al. 2013). Biological productivity is often low (typically in the order of 100 g C m^{-2} a^{-1}), and some oceanic regions like the large oceanic gyres have been described as marine deserts, largely due to severe nutrient limitation (Polovina et al. 2008). The absence of biogenic habitats is another defining feature, with the interesting exception of pelagic macroalgae (*Sargassum* sp.) floating in large mats in the Sargasso Sea (tropical Atlantic). Human influences on the pelagic ocean are certainly lower when compared to coastal regions, but still significant through the effects of fishing, pollution, shipping, and climate change (Halpern et al. 2008). Major groups of pelagic organisms include both planktonic (passively drifting) organisms and actively swimming nekton. Primary production is performed by single-celled phytoplankton, ranging from submicrometer cyanobacteria to larger diatoms and dinoflagellates, with the community composition largely determined by the nutrient regime. Likewise, the zooplankton are taxonomically diverse, ranging from single-celled protists to gelatinous "jellyfish" that may grow to over 1 m in diameter. Abundant crustaceans, such as the smaller copepods or larger euphausiids, are important food resources for higher trophic levels. Active swimmers consist of fishes, squids, cetaceans, and sea turtles. Pelagic seabirds such as shearwaters and albatrosses are also of interest, as they forage in the surface layer of pelagic waters, and spend most of their lives out above the open ocean, often returning to land only to nest.

2.2.1. Pelagic Invertebrates and Plankton

Interest in pelagic diversity patterns focused first on the Foraminifera, a class of shell-forming protozoans that deposit abundant microscopic fossils in sediments around the world. About 40 extant morphospecies are pelagic. Their microfossils were analyzed first by paleo-oceanographers, who typically used them to infer past environmental conditions from known climatic tolerances, isotope composition, and other features (Stehli et al. 1969; Berger and Parker 1970). It was noted in an early study of the North Atlantic that patterns of diversity for pelagic foraminifera were distinctly different from the "typical" pole-to-equator diversity gradient seen in coastal taxa (Ruddiman 1969). A global synthesis of available sediment core data later corroborated that the richness of planktonic foraminifera generally peaked at about 30 degrees latitude North or South, with lowest richness in high-latitude waters, and intermediate values in the tropics (Rutherford et al. 1999, and fig. 2.5). This pattern challenged the then commonly held view that species richness at a global scale tended to decline uniformly from the tropics to the poles. It further contrasted with the canonical coastal diversity pattern discussed in section 2.1, as there is no obvious longitudinal pattern to the foraminifera. Diversity in this group peaks along broad mid-latitude bands in all oceans, rather than in a distinct Indo-Pacific hotspot (Rutherford et al. 1999).

This different diversity pattern for foraminifera seems to hold for zooplankton groups more generally. Euphausiidae (krill), Pteropoda (pelagic gastropods), and Chaetognatha (arrow worms) caught in North Pacific plankton tows appear to share a similar nontropical distribution of species richness with latitude (Angel 1993, 1997). In the Atlantic, data combined from a variety of studies (Macpherson 2002) show either peaks at intermediate latitudes for a number of gelatinous zooplankton taxa (Appendicularia, Hydromedusae, Salpidae) or, alternatively, high diversity from mid- to low latitudes (Siphonophora, Chaetognatha). In stark contrast to coastal taxa, none of the groups examined showed a distinct tropical peak in diversity (Macpherson 2002).

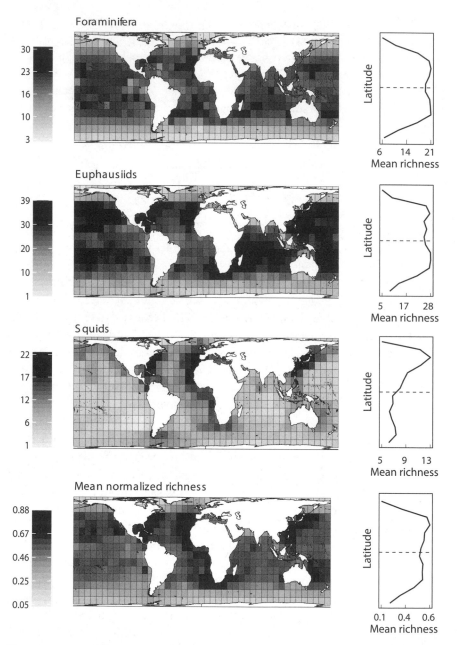

FIGURE 2.5. Pelagic invertebrates. Shown are the global species richness patterns and corresponding latitudinal gradients for pelagic zooplankton (Foraminifera and Euphausiidae), and invertebrate predators (squid). Mean normalized richness pattern averaged across these groups is shown at the bottom. After data from Tittensor et al. (2010). Note that the pattern for squid includes only commercial species.

A global distribution pattern for macrozooplankton species richness is at present available only for krill (Arthropoda; Order: Euphausiidae) (see fig. 2.5; Brinton 2000; Tittensor et al. 2010). Their pattern is similar to the foraminifera, with clear mid-latitudinal bands of high diversity that reach partly or wholly across ocean basins, and with an asymmetric pattern that showed highest richness in the southern hemisphere. The spatial correlation between the Foraminifera and Euphausiidea is high, despite significant variation in size, life history, and mobility ($r = 0.73$, $P < 0.0001$). Another key group is the copepods (Arthropoda; Subclass: Copepoda), which are a numerically dominant macrozooplankton group in many regions. Here, spatially comprehensive information is available from Continuous Plankton Recorder surveys in the Atlantic, but less so elsewhere (Beaugrand et al. 2002; Rombouts et al. 2009). There appears to be a subtropical richness peak of copepod diversity at about 20 to 30 degrees North in the North Atlantic, but no clear gradient in the South Atlantic (Rombouts et al. 2009). Note, however, that the spatial coverage of copepod data was quite limited and did not extend to higher latitudes in the South Atlantic.

Somewhat surprisingly, given their importance as the ocean's main primary producer, there is not yet a comprehensive empirical database of global phytoplankton diversity. A global modeling study, however, predicted that diversity was likely to peak at around 40 degrees latitude North and South, specifically in areas associated with western boundary currents and other transition zones (Barton et al. 2010). Similarly to zooplankton, phytoplankton diversity was predicted to be lowest at high latitudes and intermediate at the equator (Barton et al. 2010). The study used a biogeochemical model that included prognostic equations for the growth of 78 different phytoplankton types. Available empirical data support some of these patterns. For example, the diversity of coccolithophores (Class: Prymnesiophyceae) in the Pacific peaks at 30 degrees North (Honjo and Okada 1974), and the diversity of diatoms (Class: Bacillariophyta) peaks between 10 to 30 degrees North or South, depending on the size fraction examined (Malviya et al. 2016). More species-specific sampling is needed to comprehensively test model predictions (Barton et al. 2010).

Marine bacteria are a diverse group of autotrophic, heterotrophic, and mixotrophic organisms that are often considered part of the phytoplankton. Although not comprehensively sampled at a global scale, there is evidence from several independent studies for a latitudinal gradient in the diversity of marine bacteria (Pommier et al. 2007; Fuhrman et al. 2008; Amend et al. 2013; Sul et al. 2013b; Sunagawa et al. 2015). These studies challenge the commonly held view that pelagic bacteria, due to their small size and high dispersal rates, may show unique (or maybe random) patterns of diversity across large scales (Pommier et al. 2007). Interestingly, bacteria appear to have similar macroecological properties as multicellular organisms (Amend et al. 2013), and most authors described a mid-latitudinal peak in diversity (Fuhrman et al. 2008; Sul et al. 2013a; Sunagawa et al. 2015) similar to other pelagic taxa (Tittensor et al. 2010). For example, in a comprehensive metagenomics study that collected standardized microbial samples across all oceans except the Arctic, it was found that diversity of operational taxonomic units (OTUs) peaked at intermediate latitudes around 30 to 40 degrees North or South (Sunagawa et al. 2015). A geographically more limited study suggested a peak at 20 degrees North along North and Central American coastlines (Raes et al. 2011). Yet another data set suggested that microbial richness peaks seasonally at high latitudes in winter (Ladau et al. 2013). Clearly, these results are still in flux, and probably influenced in part by different sampling techniques (a common challenge for global biodiversity studies), yet all suggest a global biodiversity pattern that is more similar to other pelagic species and less similar to coastal or land species.

2.2.2. Pelagic Vertebrates and Nekton

In contrast to passively drifting and often microscopic plankton, discussed earlier, the larger nekton comprises both invertebrate and vertebrate taxa characterized by greater mobility. Pelagic cephalopods include about 300 species of squids (Mollusca; Class: Cephalopoda). Like the coastal cephalopods, spatial data for this invertebrate nektonic group are available only for commercial species, about 25% of total known richness (Tittensor et al. 2010). This might explain the highly unusual skew in peak species richness toward the northern hemisphere (see fig. 2.5), where most squid fisheries take place. Yet the general latitudinal pattern of peak richness at intermediate latitude appears probable in the southern hemisphere, and shows similarity to both the Euphausiidae and oceanic sharks (figs. 2.5 and 2.6).

As for vertebrate nekton, this comprises bony fish, sharks, rays, mammals, and seabirds. Analyses of pelagic fish diversity over biogeographic scales show patterns that quite closely resembled those seen for much smaller and taxonomically very distant pelagic organisms such as the foraminifera (Worm et al. 2003; Worm et al. 2005; Tittensor et al. 2010). Tuna and billfish (Class: Osteichthyes; Families: Scombridae, Istiophoridae, Xiphiidae) in particular are well studied due to their commercial importance, with fisheries operating worldwide for these species. Whether analyzing gridded fisheries data (Worm et al. 2005; Tittensor et al. 2010) or overlaying expert-derived range maps (see fig. 2.6), similar patterns of diversity emerge, which display a drawn-out low to mid-latitudinal peak of diversity (Worm et al. 2005), with somewhat higher average richness in the southern hemisphere (see fig. 2.6).

Pelagic sharks (Class: Chondrichthyes; Superorder: Selachimorpha; see fig. 2.6), as well as dolphins and whales (Class: Mammalia; Order: Cetacea; see fig.

2.6) again show similar patterns of mid- to high-latitude diversity, peaking around 30 to 40 degrees latitude North or South (Schipper et al. 2008; Tittensor et al. 2010; Kaschner et al. 2011; Lucifora et al. 2011), with sharks being clearly bimodal and cetaceans showing an asymmetric peak in the southern hemisphere. The pattern for sharks tends to be skewed toward the coast, whereas cetaceans appear more widely distributed throughout the oceans, and generally occur at higher latitudes than sharks. Yet it is remarkable how the most species-rich cells for both groups cluster around California, Argentina, South Africa, Japan, and Australia (see fig. 2.6), suggesting common environmental or evolutionary drivers.

Pelagic birds (Order: Procellariiformes) include the albatrosses, shearwaters, and petrels, which are uniquely adapted wide-ranging seabirds that visit the land only to nest. They show an interesting pattern of high-latitude richness, again with a distinct unimodal peak in the southern hemisphere, at around 40 degrees South (see fig. 2.6; Davies et al. 2010). The most species rich cells are found south of Australia. Davies et al. (2010) explain this pattern through the reliance of these seabirds on high and continuous wind speeds allowing long-distance travel to patchy food sources in the open ocean. Such winds are particularly found in the circumpolar waters of the Southern Ocean, where no land masses restrict the flow of air. It is in that region that pelagic seabird richness is highest. It is also noteworthy that, like the pinnipeds and cetaceans, seabirds are endotherms, which provides them with a unique competitive advantage in cold-water environments, and may in part explain their higher-than-average latitudinal distribution pattern. We further note that endothermy comes at a cost of much higher metabolic rates and calorific requirements, which may make these species seek out high-productivity regions that are typically found in higher-latitude oceans.

2.2.3. Synthesis

The combined patterns of species richness for pelagic invertebrates (see fig. 2.5) and vertebrates (see fig. 2.6) differ significantly from the patterns seen in coastal species, as well as those on land. Species richness tends to be more uniformly distributed latitudinally and longitudinally. Biodiversity hotspots are rarely found near the equator, and are typically broader and less peaked than in coastal taxa. Areas of highest richness tend to be located at intermediate or high latitudes, most commonly between 20 to 40 degrees North or South (figs. 2.5 and 2.6), with about half of the species groups being particularly diverse in the southern hemisphere (euphausiids, tunas and billfish, cetaceans, and seabirds). It is conceivable that the much larger pelagic ocean area in the southern hemisphere has promoted radiation of pelagic taxa there. Despite these hemispheric differences, there is reasonably strong overlap among spatial biodiversity patterns in the pelagic realm,

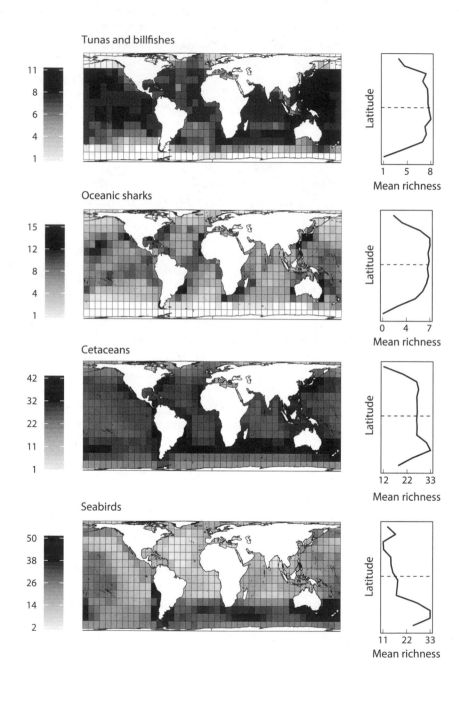

Tunas and billfishes

Oceanic sharks

Cetaceans

Seabirds

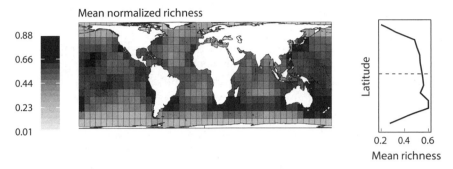

FIGURE 2.6. Pelagic vertebrates. Shown are the global species richness patterns and corresponding latitudinal gradients for tunas and billfishes, pelagic sharks, cetaceans (whales and dolphins), and seabirds (Procellariiformes). Mean normalized richness pattern averaged across these groups is shown at the bottom. After data from Davies et al. (2010); Tittensor et al. (2010); Lucifora et al. (2011).

from plankton to top predators, suggesting some common processes acting on organisms with very different modes of life, mobility, and dispersal. These environments are clearly structured by different processes compared to coastal and terrestrial environments, and the resultant patterns are distinct. Uniquely, dispersal limitation is probably relaxed in the open ocean, due to the inherent connectivity and fluidity of the environment and the high mobility of most pelagic organisms. Additionally, structured biogenic habitats are largely absent. Unlike in coastal seas, we observe consistent bands of high diversity coinciding with frontal systems that characterize the poleward boundaries of subtropical gyres, and western boundary currents such as the Gulf Stream in the Atlantic and Kuroshio in the Pacific. These transition zones are known as important habitat features that include both cold- and warm-water specialists (Block et al. 2011). Frontal systems are also known to enhance local primary production, and can attract large numbers of seabirds, marine mammals, oceanic fish, and plankton (Haney 1986; Olson et al. 1994; Etnoyer et al. 2004). Thus, these unique oceanographic habitat features may play a role in explaining the distinctive patterns of diversity seen across pelagic species groups.

2.3. DEEP-SEA BIODIVERSITY

The deep sea includes seafloor-associated (*benthic*) as well as open-water *pelagic* habitats beyond the continental shelf and slope and the shallow surface ocean (>2000 m deep). These habitats encompass the majority of the oceans' seafloor and volume, yet scientists have surveyed only a small fraction of them, due to their size and inaccessibility. The deep sea features large, seemingly monotonous plains of fine sediment dotted by intriguing anomalies like seamounts, hydrothermal vents, and cold seeps, many of which have not even been discovered or put on a map. Primary productivity is close to zero, due to the perennial darkness of the deep sea, where only some chemosynthetic organisms are able to derive energy from reducing minerals such as sulfides. Most organisms are wholly dependent on a sparse and patchy rain of detritus that slowly sinks from the sunlit surface waters to the deeper reaches of the ocean, the so-called export productivity (or "marine snow") that is not retained within the surface layers. As much of this matter is consumed on its way down, total carbon input tends to be very low, and often occurs in short and variable pulses, linked, for example to episodic plankton blooms or events such as whale-falls (Rex and Etter 2010). Another unique feature is the uniformly low temperature found in the deep sea, which typically ranges between ~2 and 4°C, with the exception of the Mediterranean deep basin. Due to both low temperature and carbon input, standing biomass and secondary production also tend to be very low, the latter often in the order of $1 \text{ g C m}^{-2} \text{ a}^{-1}$. Exceptions are the unique chemosynthetic organisms and associated fauna that colonize deep-sea hydrothermal vents and seeps, many of which are extremophiles (Rex and Etter 2010). Such habitats form productive "oases" that can sustain a high biomass of a specially adapted and unique fauna. Overall, however, the low temperature and the variable and sporadic availability of food are likely to be key organizing factors.

At the time of the *Challenger* expedition in the late nineteenth century, the deep sea was presumed lifeless due to the crushing pressure, frigid cold, absence of light, and seeming lack of productivity. Since then, this view has been overturned by the discovery of large numbers of morphologically bizarre animals and microorganisms in these remote habitats. Initially, an early extrapolation of species numbers derived from nine stations taken off the US East Coast suggested there could be many millions of undiscovered species in the deep sea (Grassle and Maciolek 1992). Although likely an overestimate (May 1992; Mora et al. 2011), that study highlighted the extreme undersampling of these habitats, and raised significant public interest in the deep sea. Since then, modest progress has been made in resolving the enigmatic patterns of deep-sea diversity. Reasonably comprehensive spatial patterns are so far available for only a few abundant and taxonomically tractable groups, and most of the patterns still rely on few samples, making it difficult to generalize (Rex and Etter 2010). Moreover, with the exception of a recent global study on brittle stars (Woolley et al. 2016), virtually all previous large-scale work has focused on the

Atlantic Ocean and has frequently examined only latitudinal gradients rather than global spatial patterns.

2.3.1. Global Patterns

The first fully global study of deep-sea spatial richness patterns focused on brittle stars (Echinodermata; Class: Ophiuroidea), which are one of the numerically dominant and most widely distributed macrofaunal taxa in the deep sea. The study used a new database comprising 165,000 species distribution records (Woolley et al. 2016). It found that deep-sea (2000–6500 m) species richness patterns fundamentally differed from those in continental shelf (0–200 m) and slope (200–2000 m) waters. While ophiuroid richness on the continental shelf and slope still matched the coastal patterns described earlier (see fig. 2.2), deep-water ophiuroid diversity peaked at higher latitudes—specifically, the highly productive waters of the northern North Atlantic, around Japan and New Zealand, and the South American and South African upwelling regions (fig. 2.7).

Specifically, as seen in fig. 2.2, shallow-water ophiuroid richness peaked in the tropics (0 to 20 degrees latitude North and South, 20–200 m; Woolley et al. 2016). A strong latitudinal biodiversity gradient exists at these depths with reduced richness at higher latitudes (>45 degrees South and >55 degrees North). However, at mid-slope to abyssal depths (200–4500 m), diversity maxima gradually shift toward temperate latitudes (30 to 40 degrees South and 40 to 50 degrees North; Woolley et al. 2016), a unique global pattern that is broadly supported by regional analyses, discussed later. Southern ocean peaks in diversity are as pronounced as

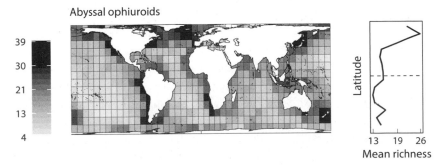

FIGURE 2.7. Deep-sea ophiuroids. Shown are the global species richness patterns and corresponding latitudinal gradients for ophiuroids (brittle stars) in waters >2000 m depth and corresponding latitudinal gradients for this species group. After data from Woolley et al. (2016).

northern hemisphere ones, but less numerous, resulting in lower average richness across a latitudinal band (see fig. 2.7). The fact that this taxon shows a very different global richness pattern in coastal and shelf seas suggests that there are unique processes structuring biodiversity in the deep sea, likely related to the patterns in the availability of energy provided by the thin rain of potential food particles descending from surface waters (Tittensor et al. 2011; McClain et al. 2012; Woolley et al. 2016). However, we caution that the general undersampling of the deep ocean and the fact that true global spatial patterns have been described only for a single taxon render our knowledge of the deep sea far more tenuous than for other major habitats; an exciting opportunity remains to further test models and theories as new data become available.

2.3.2. Regional Studies

Because of the scarcity of global richness data for deep-sea taxa, we briefly discuss relevant regional studies in addition. Latitudinal patterns in deep-sea benthic fauna (Mollusca; Classes: Gastropoda, Bivalvia) were first reported from Atlantic Ocean samples collected between 500 and 4000 m depth (Rex et al. 1993). A monotonic equator-to-poles latitudinal gradient was observed in the North Atlantic for some taxa, particularly after accounting for different sampling depths. However, this was not as clear in the South Atlantic, where highest diversity is commonly observed at high latitudes (Rex et al. 1993; Brey et al. 1994; Brandt et al. 2007). Another taxon, the sea lice (Arthropoda; Class: Isopoda), showed high diversity at both tropical and temperate latitudes in the North Atlantic, and at temperate latitudes in the South Atlantic (Rex et al. 1993). North Atlantic patterns were also influenced by anomalously low diversity data from the Norwegian Sea that probably reflect the regional effects of glaciation more than generalizable gradients of diversity (Rex and Etter 2010). No oceanwide study of these taxa has been published for the Pacific or Indian Ocean.

Small-bodied meiofauna have also been studied along regional gradients. Deep-sea benthic nematodes (Phylum: Nematoda) showed no clear patterns when plotted against latitude in the Atlantic (Lambshead et al. 2000; Rex et al. 2001), but a negative relationship appeared between 0 and 23 degrees latitude in the Pacific, possibly reflecting the positive effects of enrichment by equatorial upwelling (Lambshead et al. 2002). Likewise, benthic foraminifera (2000–4000 m depth) from the Atlantic showed a broad decline in species richness toward the poles (Culver and Buzas 2000). Peak diversity, however was observed at about 25 degrees South and 45 degrees North, possibly reflecting increased productivity at these stations. In summary, regional studies on other taxa are somewhat variable but partly correspond with the global results on ophiuroids, notwithstanding pronounced uncertainty considering limited sampling.

2.3.3. Synthesis

The vast environment of the deep sea is the most removed from the land, and has yielded the least scientific data of the four major environmental realms discussed in this book. Yet it covers 60% of the Earth's surface and appears to feature distinct patterns of biodiversity. Although most of the deep sea is perennially dark, cold, and food-limited, there are unique environments of the deep ocean—including whale-falls, seeps, and vents—that are sites of high and often unique biodiversity (Rex and Etter 2010). A global-scale study on brittle stars, supported by some regional data for other invertebrate and protozoan taxa, suggests peak diversity at higher latitudes. These zones of relatively high diversity in the abyss appear associated with areas of high surface productivity and carbon export, as well as proximity to coastlines (Woolley et al. 2016). We conclude that the lightless and cold deep sea appears to harbor unique patterns of diversity that we are only beginning to unveil.

2.4. TERRESTRIAL BIODIVERSITY

As far as we know, life in the ocean diversified for >2 billion years before the land was first colonized by plant-like "Ediacaran" life forms 635–542 Ma ago (Retallack 2013). The lack of an aqueous medium is the most distinguishing feature of the land, and has a number of important consequences for all life forms attempting to thrive there. The main obstacle, undoubtedly, is the risk of desiccation, which requires a series of unique adaptations. These include protection derived from thicker cell walls and specialized cuticular structures. Thicker cell walls also play a role in the evolution of body support that needs to be strengthened when not suspended in water. Furthermore, while algae simply absorb nutrients from the surrounding environment, a land plant must absorb both water and nutrients from the soil and transport it to its limbs. After plants began to use rigid cell walls to grow taller, a system of tubular vessels evolved to transport water and nutrients. As plants spread farther and diversified, they also began forming complex habitat structures above and below ground. These structures transformed the land surface from a largely two-dimensional to a three-dimensional habitat with many opportunities for coevolution among plants and animals. Diversification by coevolution, as well as the presence of complex structured habitats that favor reproductive isolation and speciation, might help to explain why the land over its shorter evolutionary history has produced an estimated 6.5 million species of extant eukaryotes, compared with only 2.2 million species in the ocean (Mora et al. 2011). Another possibility is that of higher evolutionary rates due to higher surface temperatures and higher levels of mutagenic UV radiation on land.

Another important difference between the ocean and the land is that nutrients are more available in most terrestrial habitats. For example, phosphorus and iron are released by weathering of rocks on land and reach the ocean only via river transport, runoff, and dust deposition. Hence, these nutrients tend to be more ubiquitous and less likely to be limiting biological productivity in terrestrial ecosystems, particularly in comparison with the open ocean and deep sea. Another contrast is that there is a vastly larger standing stock of biomass on land (~500–1000 Pg C) compared to the oceans, which feature ~2 Pg C standing biomass. This means that total nutrient utilization is much higher terrestrially, while in the ocean it is much lower, as most nutrients reside in inorganic form below the thermocline. For these reasons, rates of biological production are generally quite high on land, except in areas where low precipitation or extreme cold limit the availability of moisture. These hot or cold deserts often have productivity values comparable to the open ocean or deep sea, typically less than $100 \text{ g C m}^{-2} \text{ a}^{-1}$, while communities located in warm and moist environments often exceed $1000 \text{ g C m}^{-2} \text{ a}^{-1}$ (Whittaker 1975; Lieth and Whittaker 2012). Despite its smaller area, in total the land produces around half of the global net primary production (Field et al. 1998). In summary, the land is uniquely limited by water availability, but perhaps less strongly influenced by energy, nutrient, or food limitation than most marine environments, particularly those far from land and at great depth.

Due to the inherent bias of being a terrestrial species, our knowledge of species richness patterns on land is much more detailed than in the marine realms. Particularly for higher plants and most terrestrial vertebrates, there is now comprehensive information at a global scale (for example, Grenyer et al. 2006; Schipper et al. 2008; Jetz et al. 2009; Jetz et al. 2012a,b; IUCN 2017). Notably missing are most invertebrates, particularly insects for which no global pattern has been assembled, but see Eggleton (2000), who shows that termite diversity tends to peak in the wet tropics, along with other known groups. In contrast, the diversity of galling herbivorous insects appears to peak at 25 to 38 degrees North or South (Price et al. 1998). The absence of a more comprehensive analysis of insect diversity patterns is undoubtedly due to the massive richness of these species, rendering most of the globe severely undersampled.

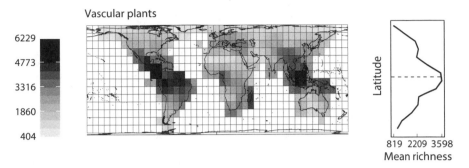

FIGURE 2.8. Land plants. Shown are the global species richness patterns and latitudinal gradients for vascular plants. After data from Kier et al. (2005); Kreft and Jetz (2007).

2.4.1. Land Plants

Comprehensive information has been compiled for vascular land plants (Tracheophyta) at a global scale (Kier et al. 2005). Generally, and with few exceptions, plants on land reach their highest diversity in the wet tropics (fig. 2.8), particularly in Southwest Asia and South America. The highest observed plant species richness is found in the Borneo lowlands (10,000 species), followed by several regions located in Central and South America with ~8000 species each (Kier et al. 2005). Diversity declines monotonically from the equator to the poles on all continents (fig. 2.8), with the exception of an extraordinary diversity of plants in a few mid-latitude hotspots such as the Mediterranean Basin and the South African Cape Floristic Region (Kreft and Jetz 2007). Dry and cold regions, such as the Gobi Desert or Greenland, show low plant diversity of <500 species per region.

Below-ground diversity on land, particularly of soil and arbuscular mycorrhizal fungi has recently been studied at some detail (Tedersoo et al. 2014; Davison et al. 2015). Broadly, these groups appear to show similar latitudinal patterns as plants, but weaker gradients toward the poles (Tedersoo et al. 2014) and much stronger overlap of species composition between continents, pointing toward their apparent ease of dispersal via miniscule spores rather than seeds (Davison et al. 2015).

2.4.2. Land Vertebrates

Land vertebrates are likely more comprehensively sampled than any other group either on land or in the sea. This detailed knowledge, and the strong overlap that

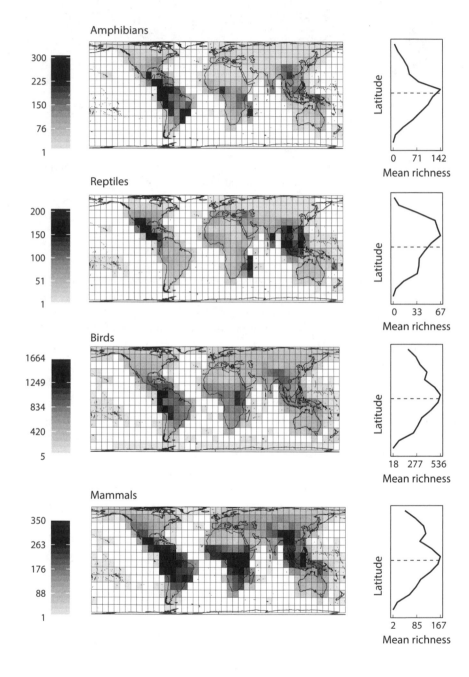

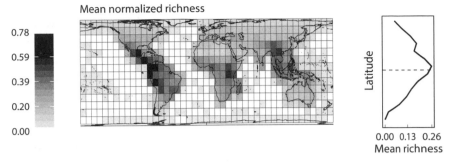

FIGURE 2.9. Land vertebrates. Shown are the global species richness patterns and corresponding latitudinal gradients for amphibians, reptiles, birds (excluding Procellariiformes, fig. 2.6), and land mammals. Mean normalized richness pattern averaged across these groups is shown at the bottom. After data from IUCN (2016). Note that all groups have >90% species coverage, except reptilians, which have about 50% coverage.

is observed between species groups, has likely contributed to the common perception of a uniform latitudinal pattern. Amphibians (Class: Amphibia), for example, show a pattern of global diversity that is exceedingly similar to plants, with major hotspots in tropical South America and Southeast Asia and rapidly declining diversity toward higher latitudes (fig. 2.9). Of course, plants and amphibians are both ectotherms and are prone to the effects of freezing and desiccation. Hence, it is maybe unsurprising that they both thrive in similarly warm and wet climates. Among the vertebrates, amphibians are likely the most susceptible to dry conditions, and uniquely among the land species groups their richness drops to zero in some of the world's driest places. Reptiles (Class: Reptilia; excluding sea turtles, sea snakes, and marine iguanas) are similar to plants and amphibians in their ectothermy, but more readily able to colonize dry places, and even deserts. Yet their diversity pattern also centers on the wet tropics, particularly Southeast Asia, but with some subtropical hotspots in Central America and Madagascar (fig. 2.9). Their latitudinal pattern peaks slightly north of the equator, reflecting the regional hotspots in Asia and Central America.

Terrestrial birds and mammals are endothermic and hence regulate their internal temperature. At the same time, they can be very mobile, in contrast to, say, plants and amphibians, and some species engage in long-distance migration. Despite these unique traits, the global richness patterns of birds and mammals are very similar to those of other vertebrates and plants (see fig. 2.9), with major peaks in the wet tropics of South America, Africa, and Asia. One major difference is the much higher diversity in Sub-Saharan Africa for birds, and particularly for mammals, when compared to plants and amphibians. It is interesting that a major contrast between endotherm and ectotherm diversity patterns is seen not on land, but in the sea. Specifically, land mammals and land birds still peak in the tropics (see fig. 2.9), whereas marine mammals and birds show their highest diversity at higher latitudes (see figs. 2.4 and 2.6).

2.4.3. Freshwater Species

Some habitats on land are moist year-round and support unique freshwater species, including plants and algae, invertebrates, fish, amphibians, birds, and mammals. Perhaps unsurprisingly, amphibians, birds, and mammals show similar geographic patterns of species richness in freshwater as they do on land, with richness patterns peaking close to the equator on all continents (Tisseuil et al. 2013). But the same is true for lake and river fishes, which may be more surprising, because one would suspect the pattern for freshwater fishes to be closer to that of their marine counterparts. Fish thus provide an interesting case study, with contrasting gradients from terrestrially influenced closed habitats (lakes) to those that are fully marine (open ocean). The similarities between freshwater fish diversity patterns and those of other terrestrial taxa may suggest common evolutionary and ecological drivers across land and freshwater habitats. Estuarine fish are partly influenced by the land, partly by the ocean. Again, perhaps unsurprisingly, their gradients in global richness appear somewhat intermediate between freshwater and coastal marine fish, but with additional variation introduced by the individual characteristics of each estuary (Vasconcelos et al. 2015). Fish diversity in coastal (see fig. 2.4) and pelagic habitats (see fig. 2.6) is again quite different from freshwater and estuarine fish, each conforming more closely to patterns seen in other, taxonomically unrelated, groups that share their habitat. This exemplifies that patterns of diversity appear much more strongly driven by the features of particular habitats (land, coast, pelagic, and deep sea) than by taxonomic affiliation.

2.4.4. Synthesis

Spatial patterns of species richness on land appear strikingly similar across plants and four different vertebrate groups (see figs. 2.8 and 2.9). This similarity occurs largely irrespective of mobility, endothermy, or trophic position (Jetz et al. 2009), and likely extends to invertebrate taxa such as insects (Eggleton 2000). Spatial correlation between most land groups is therefore high, pointing toward common environmental or evolutionary drivers. This point is reinforced by the fact that patterns of freshwater richness tend to follow those on land, at least at the broad scales examined here. The monotonic latitudinal gradient in diversity from the tropics to the poles is also clearest on land, but as we saw earlier, tends to break down progressively from coastal to pelagic to deep-sea environments, an important point that we will return to in chapter 3. One important question in paleo-ecology is whether these gradients are specific to our particular epoch, or whether they are found throughout the history of life on Earth. The next section will briefly highlight the evidence available to answer that question.

2.5. CHANGES IN BIODIVERSITY PATTERNS THROUGH TIME

Like all aspects of nature, large-scale patterns of species richness are certainly not static through time. Considerable changes in species richness may be forced by dynamic changes in climate, geological activity, plate tectonics, ocean circulation, and habitat features (Renema et al. 2008). Hotspots of benthic foraminiferan (Foraminifera living in sediments) diversity, for example, have shifted progressively over the last 50 million years from southern Europe to Southeast Asia, likely in response to tectonic changes that altered the availability of coastal shallow-water habitats on a continental scale (fig. 2.10). During the Eocene, the number of fossilized foraminiferan genera peaked in southwest Europe, northwest Africa, and along the eastern shore of the Arabian Peninsula. By the end of the Eocene, peak diversity had shifted to the Arabian Sea, and by the end of the Miocene Epoch this had shifted again toward Southeast Asia (Renema et al. 2008). This large change in the distribution of peak species richness has been discussed primarily as a consequence of coastal habitat loss largely due to regional uplift during the Arabia-Eurasia collision, likely resulting in faunal depletion and the demise of the Arabian hotspot. Fossil and molecular evidence for mollusks supports a similar pattern of "hopping hotspots" through time (Renema et al. 2008). This, and related work on corals (Leprieur et al. 2016), highlights how changes in environmental conditions and geological forces can rearrange large-scale diversity patterns over very long timescales, and emphasizes the importance of habitat availability for maintaining high regional species richness.

The observed rearrangement in benthic foraminiferan diversity over geological time entailed a shift in peak diversity from intermediate latitudes (Mediterranean) to tropical regions as the global climate cooled (note that benthic foraminifera are coastal species and as such more closely aligned with patterns in fig. 2.2 than with the pelagic foraminifera in fig. 2.5). Recently, it has been suggested that such latitudinal changes may be a more general pattern throughout the Earth's history (Mannion et al. 2014): tropical peaks in diversity tend to occur in cold "icehouse"

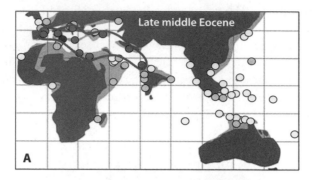

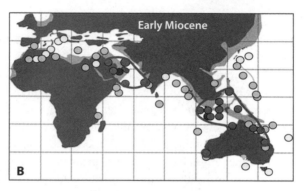

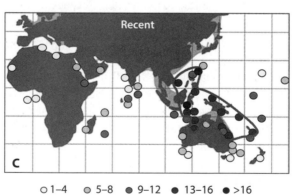

○ 1–4   ◎ 5–8   ● 9–12   ● 13–16   ● >16

FIGURE 2.10. Changes in biodiversity hotspots through time. Shown is the generic richness of large benthic foraminifera in samples from (A) the late Middle Eocene (42 to 39 Ma); (B) the Early Miocene (23 to 16 Ma); and (C) recent. Solid lines delimit the changing nature of regions that support peak diversity from the Tethyan Sea (now Mediterranean), to the Arabian Sea, to the Indonesian-Australian Archipelago (IAA). Redrawn after data from Renema et al. (2008).

climates, including the Neogene (23 Ma to present), whereas mid- to high-latitude peaks are observed both on land (for example, for dinosaurs; Mannion et al. 2012) and in the ocean (for example, for corals; Kiessling et al. 2012) during geologically extensive "greenhouse climates," which may exceed temperature maxima for many of these species (Sun et al. 2012). If confirmed, this would reveal global temperature as a major driving variable for the distribution of diversity and the shape of the latitudinal gradient. In addition, temperature-related changes in seasonality, as well as plate tectonics and mass extinction events, may further modify these patterns (Mannion et al. 2014). It should be noted that these interpretations of the fossil record rely heavily on methods used to correct for sampling biases, and are based on relatively few studies covering the "greenhouse" periods of Earth's history. Yet the evidence is mounting that tropical peaks of diversity may not have always been the norm through Earth's turbulent history and that near-lethal tropical temperatures have driven many species to higher latitudes during warmer geological periods; this may have occurred as recently as the last interglacial maximum ~125,000 years ago (Kiessling et al. 2012). In any case, these results do imply dynamic patterns of species richness through time. The question remains as to whether such dynamic changes are predictable based on what is known about the underlying drivers of diversity.

It is instructive in this regard to contrast the changes in benthic or terrestrial species richness with those of pelagic organisms. Pelagic foraminiferan microfossils are enormously abundant in deep-sea sediments, and their fossil record is probably the most complete of any marine taxon. Hence, they are particularly well understood in a paleo-oceanographic context. Studies spanning the last 3 Ma have shown that the latitudinal gradient in diversity shifted among cold and warm periods, yet the underlying relationship between diversity and temperature remained stable (Yasuhara et al. 2012). While peak diversity tended more toward the tropics in the last glacial maximum (18,000 years ago), it stretched out to higher latitudes during the last interglacial period (120,000 years ago). Similarly, the study of fossil diatom assemblages revealed a surprising consistency in community structure, and very gradual evolutionary change over the last 1.5 Ma (Cermeño and Falkowski 2009). It was shown that this apparent stability is caused by the ability of marine microbes to spatially track changes in environmental conditions, thanks to their high capacity for dispersal (Cermeño et al. 2010). As a result of relaxed dispersal limitation, the current biogeography of pelagic organisms is expected to be more closely reflective of current environmental conditions, and less shaped by past events—possibly in contrast to taxa living in coastal or terrestrial environments (Cermeño et al. 2010).

Dispersal-limited sessile organisms, such as bivalves, corals, or land plants, may require more time to adapt to large-scale changes in environmental conditions, and thus may generally show a stronger historical signature of past ecological and evolutionary factors that shaped their diversity patterns. Indeed, a spatial effect that partly captures such historic differences was significant for all land plants (Kreft and Jetz 2007), and both past and present environmental conditions have been invoked to explain contemporary patterns in terrestrial vertebrate diversity (Jetz and Fine 2012). Furthermore, in a comprehensive statistical analysis of marine richness patterns it appeared that most coastal taxa, but not a single pelagic taxon, showed a statistically significant relationship to historical differences between ocean basins (Tittensor et al. 2010). This supports the idea that dispersal constraints preserve some evolutionary footprint in sessile, or relatively slow-moving, taxa on land and in the coastal ocean.

In this context, there is considerable debate in the literature about whether present diversity hotspots for land and coastal taxa arise as centers of species origination ("cradles") or centers of species accumulation ("museums" of biodiversity; Jablonski et al. 2006). The present consensus appears to converge toward the realization that both might be true—that is, high species origination rates lead to accumulation of novel taxa in the tropics, which are largely retained there but can slowly spread to higher latitudes. This combined hypothesis has been termed the "out-of-the-tropics" model (Jablonski et al. 2006), and is well-supported by the bivalve fossil record in particular (Roy and Witman 2009). This model of species evolution assumes both higher evolutionary rates as well as lower extinction rates in the tropics (Jablonski et al. 2006; Brown 2014).

In summary, we conclude that spatial diversity patterns are not necessarily stable through time, but are influenced by historical changes in climate and habitat availability, among other factors. Such an evolutionary imprint is likely to be more visible in dispersal-limited taxa, particularly sessile coastal and land species, than in more mobile pelagic species. Clearly, it appears that the latitudinal gradient that is so often discussed in the literature is not driven by latitude per se, but by environmental factors or drivers that covary with latitude, and that these processes may vary among different habitats and throughout different periods in Earth's history. These spatial and temporal contrasts will be used to shed more light on hypothesized environmental drivers in chapter 3.

2.6. ROBUSTNESS OF DOCUMENTED BIODIVERSITY PATTERNS

The global patterns of biodiversity discussed in this section (table 2.1, later) have been derived from data collected at various levels of taxonomic resolution and using a variety of sampling methods. Here, we examine how sensitive documented diversity patterns are to changes in these two parameters. We also briefly discuss how our chosen metric of biodiversity (total species richness) relates to other metrics, which may or may not yield similar patterns at a global scale.

2.6.1. Robustness to Taxonomic Resolution

Throughout this chapter, we compiled comprehensive global patterns at the taxonomic scale that they were reported on originally, ranging from the family to the phylum level (see table 2.1). What are the effects of aggregating or disaggregating taxa on global patterns? Surprisingly, this question appears to not have been systematically explored in the literature. We investigated it for the land vertebrates, the only subphylum where we have comprehensive range information for all species (fig. 2.11). It emerges that monotonic declines in species richness from the equator to the poles are seen for the Subphylum Vertebrata, as well as for the Class Mammalia (and other classes such as Aves, Amphibia; see fig. 2.9). The Order Carnivora (fig. 2.11) still shows a clear gradient, but this becomes a lot more variable as one disaggregates to finer taxonomic resolution. At the family level, the Ursidae (bears) and Felidae (cats) both show idiosyncratic patterns that are distinct from other vertebrates or mammals (fig. 2.11), and are likely shaped by their evolutionary history and unique adaptations. Bears show their greatest richness in the temperate northern hemisphere, whereas cats peak in the subtropics in both hemispheres. We conclude that at least for the vertebrates, latitudinal gradients are stable from the class level and up, and start to break down at lower levels in the taxonomic hierarchy. One possible explanation for this pattern according to the out-of-the-tropics model discussed earlier (Jablonski et al. 2006) is that most phyla and classes, and possibly orders, may have originated in warm climates, and later spread to higher latitudes, with new families, genera, and species originating there, giving the possibility of differing latitudinal distributions for such taxa. Generally, we urge the reader to bear taxonomic resolution in mind when interpreting patterns of diversity with latitude.

2.6.2. Robustness to Sampling Methodology

Possibly, some of the variation seen between groups and habitats could be driven by differences in sampling and approaches to producing data, rather than environmental or evolutionary drivers. Broadly, there are three fundamentally different methods with slightly different limitations and subsequent inference: (1) Biodiversity patterns may be constructed by overlaying individual species ranges.

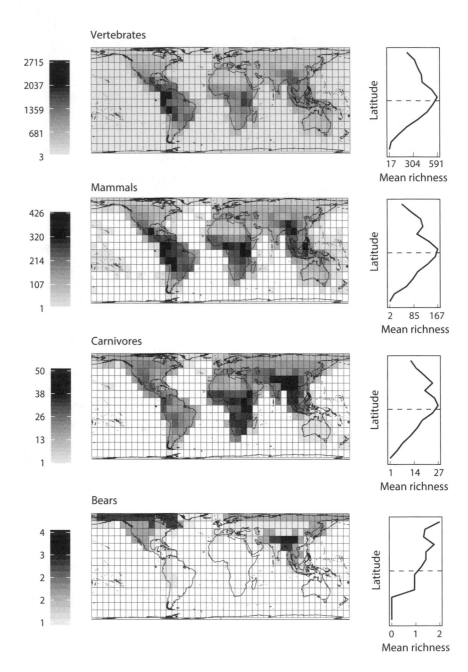

Vertebrates

2715
2037
1359
681
3

Latitude

17   304   591
Mean richness

Mammals

426
320
214
107
1

Latitude

2   85   167
Mean richness

Carnivores

50
38
26
13
1

Latitude

1   14   27
Mean richness

Bears

4
3
2
2
1

Latitude

0   1   2
Mean richness

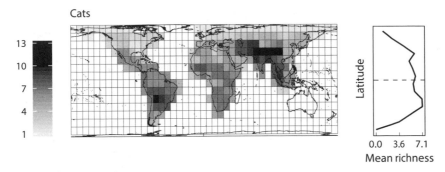

FIGURE 2.11. Sensitivity of vertebrate diversity patterns to changes in taxonomic resolution. Shown are subsets of the subphylum Vertebrata—specifically, the Class Mammalia, which contains the Order Carnivora, which in turn contains the families Ursidae (bears) and Felidae (cats), among others. It is evident how the clear latitudinal gradient for the vertebrates becomes more idiosyncratic when splitting data into progressively smaller taxonomic units. Each step down the taxonomic hierarchy reduces the number of species by about one order of magnitude.

These ranges are typically based on observations of species presence and some degree of interpolation between sampling stations, often informed by expert opinion, to estimate the total extent of occurrence. This method is commonly used for paleontological data, but also to map the range of total presence in extant species. It provides a cumulative view of biodiversity, and can overestimate actual ranges (as there are likely to be gaps within bounding polygons), and hence local biodiversity. (2) A second approach is to construct sample-based biodiversity patterns based on actual biodiversity surveys where the occurrences of all species in the community are recorded together, and then spatially interpolated by geostatistical methods such as kriging or relating to environmental parameters. This method is likely to underestimate biodiversity if sampling is incomplete or accumulation curves uncorrected (though they can be and sometimes are extrapolated toward asymptotes using various methods). This approach is also unlikely to fully represent rare species, and does not provide species identities at interpolated sites. (3) A third approach is to model individual species ranges using environmental (and possibly other) predictors—that is, to use habitat or niche models, and then overlay these species ranges to reconstruct aggregate biodiversity patterns. This method may overestimate range sizes, and hence biodiversity, if other factors (such as biotic interactions or historical disturbances) cause marked differences between realized and fundamental niches and hence restrict actual species ranges within a possible environmental envelope. Potentially, it may also underestimate range sizes if the niche is not adequately sampled and characterized.

Coastal fishes are one group for which all of these approaches have been implemented at a global scale. Reassuringly, the resulting patterns are very similar, suggesting that the basic features of large-scale biodiversity patterns are captured independently of sampling method (fig. 2.12). Whether modeling global occurrence records (Tittensor et al. 2010) or compiling transect-based samples (Edgar et al. 2014), expert-derived range maps (Roberts et al. 2002), or habitat models (Selig et al. 2014), all observed patterns centered around a primary diversity hotspot in the western tropical Pacific, with richness decreasing latitudinally and to a lesser degree longitudinally from there. Likewise, fish richness in the Atlantic centered on the Caribbean, irrespective of sampling method (fig. 2.12). This suggests some degree of robustness to differences in sampling methods and data processing for derived measures of species richness, at least at the coarse scales we choose to examine here. However, how does species richness relate to other measures of diversity, and is it a reasonable proxy for functional diversity, evenness, or phylogenetic diversity? We explore this question briefly in the following section.

2.6.3. Relationship to Other Diversity Metrics

Although in this book we maintain a deliberate focus on large-scale patterns of species richness, it is interesting to review briefly what is known about other measures of global-scale diversity, and how they relate to species richness. Of much recent interest is functional richness, which captures the number and sometimes also the relative abundance of unique functional groups or "guilds" in an area. This measure avoids the "ecological redundancy" that may be present in speciose but functionally poor communities with large numbers of similar species—although note that functional redundancy can still be a useful property in a dynamic environment (Elmqvist et al. 2003). However, functional groups can be constructed and assigned in many different ways, possibly leading to different patterns. Regardless of these complexities, it appears that gradients in species richness broadly translate into similar gradients of functional richness (Micheli and Halpern 2005; Berke et al. 2014). A detailed analysis of >5000 coastal bivalve taxa functionally categorized species according to their feeding mode, relationship with the substratum, mechanism of attachment, mobility, and body size (Berke et al. 2014). Interestingly, only 39 out of 197 possible trait combinations were realized in nature. These 39 observed functional groups were distributed nonrandomly along the world's coastlines, with low-latitude hotspots in the tropical Indo-Pacific and Eastern Pacific-Caribbean, respectively, as seen for the species richness of mollusks (see fig. 2.1) and coastal species more broadly (see figs. 2.2 to 2.4). The authors found a saturating relationship between functional

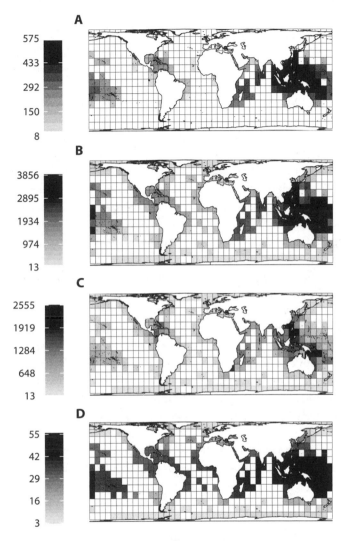

FIGURE 2.12. Sensitivity of fish diversity patterns to different sampling methods. Shown are patterns of species richness based on (A) expert-derived range maps for reef fishes (after data in Roberts et al. 2002), (B) habitat-model derived range maps for all fishes (after data from Selig et al. 2014, updated October 2017 using www.aquamaps.org), (C) extrapolated occurrence records for coastal fishes (after data from Tittensor et al. 2010), and (D) transect sampling of reef fishes (mean richness per transect) by divers (after data from Edgar et al. 2014). Data sources and methods for estimating species richness were completely independent, yet show very consistent patterns at the global scale. Data have been restricted to coastal regions where necessary.

richness and genus richness that has also been described for land taxa such as mammals (Safi et al. 2011). Likewise, the pattern of functional richness in coastal fishes closely resembles the overall pattern of species richness at global (Stuart-Smith et al. 2013) and regional scales (Micheli and Halpern 2005). These results suggest a direct although nonlinear link between the number of species and functions in an ecosystem.

In contrast to functional richness, bivalve functional evenness displays a reverse pattern, peaking at higher latitudes (Berke et al. 2014). Similarly, species evenness of reef fishes tended to be higher at higher latitudes, and lowest in the tropics (Stuart-Smith et al. 2013). At least for bivalves, which have a detailed fossil record available, this contrasting pattern of richness and evenness was explained by uneven origination rates of different functional groups in the tropics, and the movement of a small, random subset to higher latitudes (Jablonski et al. 2006). Simulation models showed that such processes could lead to higher evenness at high latitudes, indirectly supporting the out-of-the-tropics model for the analyzed bivalve communities (Berke et al. 2014). Data were broadly consistent with the hypothesis that high-latitude fauna is to a first approximation an attenuated sample of the global species pool.

Finally, there has been considerable and growing interest in patterns of phylogenetic richness, which describes the number of unique evolutionary lineages in a species group. This metric reflects the recognition that species are not independent entities, but rather their functional and ecological similarities are shaped by patterns of common ancestry (Harvey and Pagel 1991); it also has conservation implications in terms of preserving evolutionary distinctness. The idea is that it is not necessarily the most species-rich taxa that harbor the greatest diversity of lineages, particularly if one or a few groups have radiated profusely but with minimal evolutionary novelty. As with functional richness, there is a problem with lumping or splitting lineages, and no simple "standard" measure is available at this point. Still it appears that at large scales, phylogenetic richness does follow the pattern of species richness surprisingly well, for example, in mammals (Safi et al. 2011). Taken together, these results suggest some support for the use of species richness as a surrogate of both functional and evolutionary complexity for some taxa, though it may not be reflective of other measures such as evenness.

2.7. SYNTHESIS

Our comparative analysis of global patterns in species richness yielded some interesting generalizations. Averaging across all known species groups on land and in the sea, tropical peaks in species richness were as common as subtropical peaks ($n = 16$ each; see table 2.1 and fig. 2.13), whereas species groups cresting in temperate ($n = 6$) or polar latitudes ($n = 2$) were more exceptional. Thus the oft-cited unimodal pattern of biodiversity appears frequently, particularly on land, but there is also evidence that supports a newly emerging paradigm of asymmetric unimodal or bimodal peaks, often in the subtropics, and particularly in the marine realm (Chaudhary et al. 2016). We will return to this difference in marine and terrestrial gradients in the next chapter, on drivers. Longitudinally, a clear global maximum in the eastern hemisphere (mostly in South Asia) was seen in 10 groups, whereas global richness peaked in the western hemisphere (Americas) in only 2 taxa. Most groups ($n = 20$), however, showed maxima in species richness in both hemispheres (see table 2.1). These broad patterns across habitats were robust to differences in taxonomic resolution across species group: when all taxa below the class level were excluded, the overall latitudinal richness pattern remained similar (fig. 2.13).

The present synthesis (see table 2.1 and fig. 2.13), confirms earlier studies of contemporary and fossil-derived patterns of species richness in the sea, indicating flatter latitudinal gradients and bimodal latitudinal richness patterns, especially for pelagic taxa (Tittensor et al. 2010; Powell et al. 2012; Chaudhary et al. 2016). Yet this is the first globally comprehensive comparison of diversity patterns on land and in all three major marine realms. While diversity patterns on land (including freshwater habitats) tended to converge on one broad unimodal pattern with a strong tropical peak, marine biodiversity patterns were dissimilar from the land and from each other—coastal, pelagic, and deep-sea taxa showed a progressively greater tendency for peak diversity at higher latitudes (see fig. 2.13), and also displayed dissimilar longitudinal patterns (see table 2.1). As the examples of fishes, mammals, and birds indicate, processes affecting diversity tend to vary more by realm (land, coastal, pelagic, or deep sea) and less by taxonomic group—that is, diversity patterns are more similar between different taxa in the same environmental realm than within a single taxon that occurs across different realms.

Total known eukaryotic species richness for just those taxa sampled and presented in table 2.1 is likely highest on land (~325,000), intermediate in the coastal ocean (~22,000 species plus ~1000 macroalgal genera), and lowest in the pelagic realm (~500 species)—though this of course does not reflect total richness across these taxa, as it is biased toward species we sample well, and also ignores the raft of taxa for which global patterns remain undescribed. Species richness in the deep sea is too poorly sampled (a single taxon) to directly compare at a global scale. This contrast in absolute richness has not been explained comprehensively, but probably has to do with the presence of structured habitats on land and in coastal regions. Such features pose constraints on dispersal and favor speciation

TABLE 2.1. Synthesis of Global Biodiversity Patterns

Habitat	Species group	Taxonomy	Species	Latitudinal peak	Longitudinal peak	Source
Coastal	Stony corals	Order: Scleractinia	794	Tropical	East Asia	Tittensor et al. 2010
Coastal	Brittle stars	Class: Ophiuroidea	126	Tropical	East Asia	Woolley et al. 2016
Coastal	Bivalves	Class: Bivalvia	~10,000	Tropical	East Asia	Valentine and Jablonski 2015
Coastal	Cone snails	Family: Conidae	632	Tropical	East Asia	IUCN 2016
Coastal	Cephalopods*	Class: Cephalopoda	122	Subtropical	East Asia	Tittensor et al. 2010
Coastal	Mangroves	Division: Tracheophyta (5 Families)	32	Tropical	East Asia	Tittensor et al. 2010
Coastal	Seagrasses	Order: Alismatales (4 Families)	60	Subtropical	East Asia	Tittensor et al. 2010
Coastal	Macroalgae	Division: Chlorophyta	116 (genus)	Tropical	Various	Keith et al. 2014
Coastal	Macroalgae	Division: Rhodophyta	682 (genus)	Subtropical	Various	Keith et al. 2014
Coastal	Macroalgae	Class: Phaeophyceae	252 (genus)	Temperate	Various	Keith et al. 2014
Coastal	Fish	Class: Osteichthyes	9713	Tropical	East Asia	Tittensor et al. 2010
Coastal	Sharks	Superorder: Selachimorpha	480	Subtropical	East Asia	Tittensor et al. 2010
Coastal	Pinnipeds	Suborder: Pinnipedia	36	Polar	Various	Tittensor et al. 2010
Pelagic	Foraminifera	Class: Foraminifera	88	Subtropical	Various	Rutherford et al. 1999
Pelagic	Copepods*	Subclass: Copepoda	NA	Subtropical	NA	Rombouts et al. 2009
Pelagic	Euphausiids	Order: Euphausiacea	100	Subtropical	Various	Tittensor et al. 2010
Pelagic	Squids*	Order: Teuthida	85	Temperate	Various	Tittensor et al. 2010
Pelagic	Tuna and billfish	Suborder: Scombroidei (3 Families)	12	Tropical-subtropical	Various	Tittensor et al. 2010
Pelagic	Sharks	Superorder: Selachimorpha	27	Subtropical	East Asia	Tittensor et al. 2010
Pelagic	Cetaceans	Infraorder: Cetacea	81	Temperate	Various	Tittensor et al. 2010
Pelagic	Seabirds	Family: Procellariiformes	110	Temperate-polar	Various	Davies et al. 2010
Pelagic	Bacterioplankton*	Phyla (various)	562 (OTU)	Subtropical	NA	Pommier et al. 2007

Habitat	Species group	Taxonomy	Species	Latitudinal peak	Longitudinal peak	Source
Pelagic	Bacterioplankton*	Phyla (various)	>1000 (OTU)	Subtropical	NA	Fuhrman et al. 2008
Pelagic	Microbial plankton*	Phyla (various)	35,000 (OTU)	Subtropical-temperate	NA	Sunagawa et al. 2015
Deep sea	Brittle stars	Class: Ophiuroidea	31	Temperate	Various	Woolley et al. 2016
Land	Vascular plants	Division: Tracheophyta	>200,000	Tropical	Various	Kreft and Jetz 2007
Land	Soil fungi*	Phyla (various)	94,255 (OTU)	Tropical	South America, Asia	Tedersoo et al. 2014
Land	Amphibians	Class: Amphibia	6475	Tropical	South America	Grenyer et al. 2006
Land	Reptiles*	Class: Reptilea	3753	Subtropical	Central America, Africa, Asia	IUCN 2016
Land	Birds	Class: Aves	10,423	Tropical	South America, Asia	Grenyer et al. 2006; IUCN 2016
Land	Mammals	Class: Mammalia	5266	Tropical	South America, Africa, Asia	Schipper et al. 2008; IUCN 2016
Freshwater	Fish	Class: Osteichthyes	8870	Tropical-subtropical	Americas, Africa, Asia	Tisseuil et al. 2013
Freshwater	Amphibians	Class: Amphibia	3263	Tropical-subtropical	Americas, Africa, Asia	Tisseuil et al. 2013
Freshwater	Birds	Class: Aves	699	Tropical	South America, Africa, Asia	Tisseuil et al. 2013
Freshwater	Mammals	Class: Mammalia	119	Tropical	South America, Africa, Asia	Tisseuil et al. 2013
Freshwater	Crayfish	Superfamily: Astacoidea	462	Subtropical	North America	Tisseuil et al. 2013
				Latitudinal peak	Longitudinal peak	
				Tropical 16	Western Hemisphere 2	
				Subtropical 16	Eastern Hemisphere 10	
				Temperate 6	Both 20	
				Polar 2		

Note: Shown are the approximate climatic zones where species richness peaks for different species groups. Most groups have comprehensive global coverage; those with more limited global sampling or limited species coverage are marked with an asterisk (*). Most studies reported richness of species, except for macroalgae (richness of genera), and bacterioplankton and soil fungi (richness of operational taxonomic units, OTU). Latitudinal and longitudinal patterns of peak richness are summarized at the bottom; see also fig. 2.13 for a graphical summary.

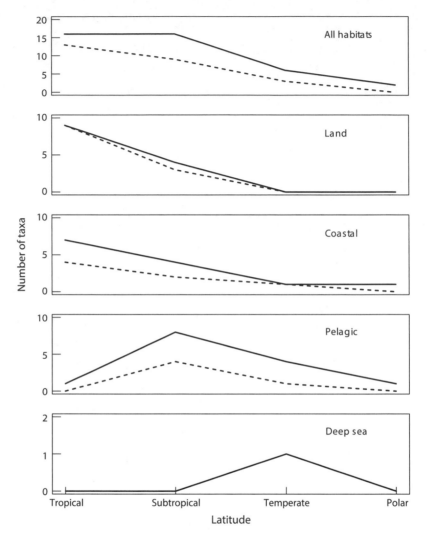

FIGURE 2.13. Synthesis of latitudinal biodiversity patterns. Shown are the approximate locations of latitudinal peaks in species richness in all habitats combined, and separated. Solid line: all taxa with global data; dashed line: all taxa below Class level removed. Data from table 2.1.

by isolation, seen for example around mountain chains on land and in the coastal maze of the Indonesian-Australian Archipelago (Bellwood et al. 2012). Land and coastal habitats also tend to be the most productive, with significant nutrient sources and frequent vertical mixing. Productivity drops off with increasing distance from shore and increasing water depth, and is lowest in the deep sea, which receives only a fraction of the surface productivity from sunlit pelagic waters. At the same time, the impacts that people have on marine biodiversity tend to attenuate with increasing distance from land (Halpern et al. 2008) and increasing depth (Pauly et al. 2003)—though an increasing human footprint is now visible even in the deep sea (Ramirez-Llodra et al. 2011; Levin and Le Bris 2015; Jamieson et al. 2017).

Some species move between habitats during their life history, but most are primarily associated with one of the four major realms discussed here, rendering the unique patterns of richness found within each realm largely independent. Moreover, as we will see in the next chapter, the environmental predictors and potential drivers of diversity are often spatially distinct between major environmental realms. As such, the contrasting biodiversity patterns among land, coastal, pelagic, and deep-sea habitats offer a powerful contrast and rich testing ground for ideas about the fundamental structuring forces of diversity at global scales, and how they may play out over evolutionary and ecological time. We will examine empirical evidence for those possible structuring forces in detail in chapter 3, and then develop a general theory of the processes that may connect diverging patterns and drivers in chapter 4.

CHAPTER THREE

Drivers and Predictors
of Global Biodiversity

Whereas we discussed global patterns of biodiversity in the preceding chapter, we now turn our attention to the driving factors that may cause these patterns to exist. As the distribution of diversity is highly nonrandom, and often similar across disparate taxa (chapter 2), there has been extensive speculation about underlying drivers and mechanisms that may influence species richness. Most of this discussion has taken place with respect to the strong latitudinal gradients of diversity found in terrestrial systems. Many (some would say too many) hypotheses have been proposed to explain these gradients, and there is little general consensus on their relative merit (Fischer 1960; Pianka 1966; Rohde 1992; Brown 2014). Most authors tend to agree, however, that a consistent pattern that emerges across taxa, such as the inverse relationship between latitude and richness on land, requires a general mechanism, or group of mechanisms, that operates independently of the particular habitat or species group examined (Rohde 1992; Gotelli et al. 2009). In the following, we attempt to gain better understanding by examining hypothesized *drivers* (factors that may mechanistically explain changes in diversity) and relating them to *predictors* (environmental variables that can be empirically measured and correlated with observed diversity patterns, then related back to drivers). It is our conjecture that a comparative global analysis of environmental predictors across land and sea could shed light on some of the underlying drivers and mechanisms that may operate within and across realms.

A fundamental challenge for such an analysis lies in the fact that many environmental predictors covary with latitude, such as annual mean temperature, temperature range, solar radiation, seasonal climatic variability, soil moisture, and the intensity of historical disturbance caused by ice ages, among others. Hence the more traditional approach of analyzing a simple, linear gradient of species richness over latitude may not be suitable to resolve conflicting hypotheses as to which drivers are causing these gradients. A comprehensive global spatial analysis is more promising, because it includes longitudinal as well as latitudinal variation in these predictors across different continents (or ocean basins) and thus provides significantly more contrast to separate their relative influence (Kreft and Jetz 2007; Tittensor et al. 2010; Jetz and Fine 2012). Such an analysis has the potential to be particularly powerful when analyzing multiple terrestrial and marine taxa, as they can show contrasting patterns of diversity (see chapter 2), and feature different environmental gradients and drivers, hopefully shedding light on underlying unifying principles. The goal of this chapter is to confront published hypotheses about putative drivers of diversity with comprehensive empirical information on the environmental predictors of diversity on land and in the oceans. Ultimately, we hope to identify common drivers and mechanisms that could form the basis for a synthetic theory of global biodiversity patterns.

3.1. HYPOTHESIZED DRIVERS OF DIVERSITY

Many hypotheses have been formulated to identify drivers and describe associated mechanisms that may influence large-scale patterns of species richness (table 3.1). Conceptually, these hypotheses fall into three broad categories: (1) those concerned with environmental factors that are thought to promote diversity (such as solar radiation, thermal energy, productivity, and environmental stability); (2) those concerned with factors that may limit diversity (such as environmental stress or disturbance); and (3) those that focus on size and quality of the habitat, within which observed diversity patterns emerge (table 3.1). It is remarkable, however, that as yet there is no body of theory that combines these three dimensions of environmental variation to explain, in a testable framework, biodiversity patterns at the planetary scale.

We note here that hypothesized drivers from the three preceding categories can operate to promote or limit diversity via the same mechanism—for example, increased energy availability and increased habitat area can both result in a larger community, which, all else being equal, can increase total diversity (the *more individuals hypothesis*; Wright 1983; Hurlburt and Jetz 2010). This hypothesized relationship between a higher number of individuals and greater diversity can itself operate through multiple pathways, such as increased sampling of the regional species pool, or reduced extinction rates due to larger population sizes. In this chapter, we focus primarily on the hypothesized drivers of diversity and their relationship to environmental correlates, but we discuss potential mechanisms wherever possible, to explore them in more detail in our model simulations in chapter 4. Note also that one might separate out potential drivers, as they may scale differently even if they operate through the same mechanism (Hurlburt and Jetz 2010). We explore such scaling relationships when fitting our models to observed data in chapter 5.

TABLE 3.1. Hypothesized Drivers and Predictors of Diversity

Hypothesized driver	Hypothesized mechanism	Scale	Environmental predictor	Source
Drivers promoting diversity				
Solar energy	More individuals can coexist.	Ecological	Solar insolation	Currie 1991
Thermal energy	Higher community turnover and speciation rate.	Evolutionary	Surface temperature	Brown et al. 2004
Productivity	More individuals can coexist.	Ecological	Net primary production	Wright 1983
Environmental stability	Lower extinction rate.	Evolutionary	NA	Sanders 1968
Evolutionary time	More niches get filled over time.	Evolutionary	Years	Fischer 1960
Drivers limiting diversity				
Environmental stress	Fewer species can adapt to stressful condition.	Ecological	Various	Thiery 1982
Disturbance	Higher extinction rate.	Evolutionary	Various	Fischer 1960
Seasonality	Unstable environment limits specialist niches.	Ecological	Variation in net primary production	Taylor and Taylor 1977
Drivers relating to habitat				
Habitat size	More individuals can coexist.	Ecological	Area	Connor and McCoy 1979
Habitat complexity	More niches are available.	Ecological	Various	Pianka 1966
Habitat shape	Species richness peaks in mid-domain.	Ecological	NA	Colwell & Lees 2000

Note: Only major hypotheses that are thought to apply across taxa and are considered in multiple publications are listed (source publications serve as examples). Mechanisms and predictors refer to the most common interpretation; alternative mechanism and predictors are possible.

Finally, we note that we do not attempt a complete survey of all mechanisms associated with each driver. For example, Evans et al. (2005) list nine potential mechanisms that may generate positive relationships between ambient energy availability and species richness. We focus on those mechanisms that have received the broadest support within the literature, and note that other factors may well play a role. A comprehensive test of all alternative mechanisms within our analytical framework will be possible, but will not be attempted here.

3.1.1. Drivers Promoting Diversity

Life on Earth requires external energy input from the sun, as well as a set of essential resources such as carbon, water, and nutrients. A number of authors have proposed that large-scale patterns of species diversity and the latitudinal gradient on land, in particular, may be generated and maintained by greater energy availability toward the equator. The concept of biologically available energy promoting diversity, whether such energy is in the form of photosynthetically active solar radiation (PAR), thermal energy, or chemical energy (Gibbs free energy in the tissues of organisms), has been formalized in the *species-energy theory* (Wright 1983; Clarke and Gaston 2006). Hypothesized mechanisms by which energy may affect diversity fall into two broad classes: (1) Increased energy availability leads to greater diversification due to increased metabolic rates, faster community turnover, and/or more rapid evolutionary processes; this is the *evolutionary-speed hypothesis* (Stehli et al. 1969; Rohde 1992; Allen et al. 2002). (2) Increased energy availability may support higher biological productivity and sustain more individuals per unit area, which allows more species to coexist; this is the *more-individuals hypothesis* (Hutchinson 1959; Wright 1983; Clarke and Gaston 2006). Note that these mechanisms are of course nonexclusive; they could both operate simultaneously, the former on evolutionary and the latter on ecological timescales (see table 3.1). Here, we examine the three forms of energy that may promote diversity—solar, thermal, and chemical—followed by an examination of the influence of environmental stability over time in promoting diversity.

SOLAR ENERGY

There is marked latitudinal variation in the seasonality and intensity of solar energy input, the latter due to the shallower angle of incidence and greater scattering in the longer atmosphere path length toward higher latitudes. When averaged over the year, the difference between received solar energy (in $W\ m^{-2}\ a^{-1}$) at the tropics and poles is about fourfold (Öpik and Rolfe 2005). Where PAR, water, and sources of nutrients are abundant, plant productivity may be elevated, which channels more available energy to higher trophic levels.

Conversely, lower levels of PAR might lead to lower plant diversity because, all else being equal, fewer individuals can coexist on a given amount of incoming energy; this is the more-individuals hypothesis described earlier (Hutchinson 1959; Wright 1983; Evans et al. 2005). However, only a small fraction of the incident PAR, typically less than 1%, is used by photosynthetic organisms (Öpik and Rolfe 2005); thus, incident PAR itself may not typically be a strong limiting factor that drives diversity gradients. Furthermore, such differences in PAR cannot explain the altitudinal diversity gradient. As is well established on land, diversity decreases for most species groups with increasing altitude (Gaston 2000), and microevolutionary rates appear lower at higher altitudes (Bleiweiss 1998; Gillman et al. 2009). Yet solar energy input is constant, or even increasing due to shorter atmospheric pathlength, with increasing altitude. Finally, a direct effect of solar irradiance on species richness is not compatible with the extratropical richness peaks commonly seen in pelagic organisms (see chapter 2). A mechanism by which solar energy could affect diversity indirectly, however, is through its effects on (1) surface temperature or (2) productivity. These mechanisms will be discussed in the two following sections.

SURFACE TEMPERATURE

Elevated solar energy input into tropical environments not only affects PAR but also elevates average surface temperature compared to temperate or polar areas. On first glance, it appears obvious for most taxa that more species prefer warm conditions to cold ones; hence temperature tolerances could explain gradients of biodiversity for many species groups along both latitudinal and altitudinal clines (Boyce et al. 2008; Sunday et al. 2011; Beaugrand et al. 2013). Yet, like other explanations based on the concept of perceived environmental harshness and tolerance, this is ultimately circular reasoning, as "thinking of the tropics as benign and the polar region as harsh is only a habit of thought; it results from the fact that life is more abundant in the tropics" (Pielou 1979; see also the discussion in section 3.1.2). More specifically, it raises the key question of *why* more species evolved in warmer environments in the first place, and the mechanisms underlying this relationship.

A number of hypotheses have been proposed that link higher temperature to a faster speed of evolution (the aforementioned evolutionary-speed hypothesis), hence providing an evolutionary mechanism linking (thermal) energy to diversity (Clarke and Gaston 2006). Prominent among these hypotheses is the *metabolic theory of ecology* (Brown et al. 2004), which posits that the metabolic rate is the fundamental biological rate that governs many macroecological patterns in nature. The theory is based on empirically observed relationships between body size, temperature, and metabolic rate across all organisms. Specifically, small-bodied organisms tend to have higher mass-specific metabolic rates than larger-bodied ones. This relationship is empirically described by Kleiber's law,

$$B = B_0 M^{3/4},$$

where B is the whole organism metabolic rate (in watts or another unit of power), M is organism mass (in kg), and B_0 is a mass-independent normalization constant. The constant exponent of 0.75 (3/4) is thought to be explained by fractal scaling of resource distribution networks (West et al. 1997).

Furthermore, organisms that operate at higher temperature (either through endothermy or by living in warm environments) have higher metabolic rates than those at colder temperatures. This fundamental temperature dependence of biological reactions is described by the Arrhenius function,

$$I = ae^{-E/kT},$$

where I is a physiological rate, a is an intercept, E is the activation energy in electron volts (typically around 0.6–0.7 eV for respiration), k is the Boltzmann constant (8.62×10^{-5} eV K^{-1}), and T is the internal temperature of the organism in degrees Kelvin (K). Note, however, that the assumed log-linear temperature dependence holds up only to a species-specific maximum temperature, above which metabolic rates decline quickly due to the denaturation of proteins at high temperature (Corkrey et al. 2014). Kleiber's law and the Arrhenius function were unified into a metabolic theory of ecology (Gillooly et al. 2001), which modeled metabolic rates as a function of both body size and temperature, following

$$B = B_0 M^{3/4} e^{-E/kT}.$$

The main implication is that metabolic rate, as influenced by body size and temperature, provides the fundamental constraint by which ecological processes are governed. If this holds true, many ecological patterns from the level of the individual up to ecosystems might be explained, at least in part, by the relationship between metabolic rate, body size, and temperature.

The theory is supported by the finding that underlying physiological mechanisms are very general. Recently, it was shown that the same metabolic temperature dependence function applies to prokaryotes as well as unicellular and multicellular eukaryotes, from extreme-cold-water-adapted to hyperthermophilic forms (−2 to 120°C) (Corkrey et al. 2014). Apparently, the fundamental kinetics remain the same across the whole range of known thermal adaptations and can actually be used to predict protein thermodynamics directly from growth rate data. This means that all known life forms, which evolved at different times over at least 3 billion years, can be at least broadly described by the same temperature-dependence model, implying a single, highly conserved reaction that may well trace back to the last common ancestor of prokaryotes and eukaryotes. The generality of such molecular mechanisms supports, at least in principle, the metabolic theory (Corkrey et al. 2014), and may in part explain why temperature effects on many ecological patterns and processes including biodiversity are as reproducible as they are.

In light of such generality, the domain of metabolic theory has been gradually extended beyond the original focus on organismal biology. Most relevant in the context of this book is the *metabolic theory of biodiversity* (Allen and Gillooly 2006), which is an extension of the metabolic theory of ecology. It assumes that both the number of generations per unit time (community turnover) and the number of DNA mutations increase with increasing metabolic rate, and thus with temperature. It further assumes that the rate of speciation is proportional to mutation rate. Combining these assumptions leads to the prediction that speciation rate, and (by inference) species richness, increase exponentially with temperature (Allen et al. 2006).

Assumptions and predictions generated by the metabolic theory of biodiversity have been tested empirically, with varying results. On the one hand, there is some good evidence that the "molecular clock" of nucleotide substitution and DNA evolution (Gillooly et al. 2005), as well as rates of genetic divergence and speciation over geological time (Allen et al. 2006), are positively related to temperature (fig. 3.1A). There is also some detailed experimental evidence that shows increased mutation rates at higher temperature (Muller 1928; Lindgren 1972), particularly when ambient temperature approaches the upper tolerance limit of a species, inducing physiological stress (Matsuba et al. 2013). Temperature stress may directly interfere with DNA repair mechanisms, which are more error-prone when individuals are in poor condition (Agrawal and Wang 2008). Alternatively, the *metabolic-rate hypothesis* posits that most mutations are caused by genetic damage from free radicals produced as by-products of metabolism; therefore, mutation rate should be related to mass-specific metabolic rate and hence body temperature (Martin and Palumbi 1993).

Species richness, on the other hand, does not always display the straightforward relationship with temperature predicted by metabolic theory (Algar et al. 2007; Hawkins et al. 2007). It has been hypothesized that other factors, like water availability (Hawkins et al. 2003), may confound or override the influence of temperature on species richness. Note, however, that this specific factor would not apply to the marine environment. Generally speaking, while increased mutation rate under elevated temperature has been demonstrated many times, the link to speciation and macroevolutionary rates has not been established experimentally, and the mechanisms remain unclear.

Empirically, however, the fossil record provides compelling correlative evidence that such a link may exist. Long before the metabolic theory of biodiversity was proposed, a relationship between temperature and speciation rate had already been inferred from the marine fossil record, particularly for mollusks and corals, where evolutionary rates for warm-water fauna seemed to exceed those of cosmopolitan or cold-water species, leading to the typical equator-to-poles richness gradient over latitude. This gradient was seen in fossil marine assemblages at least

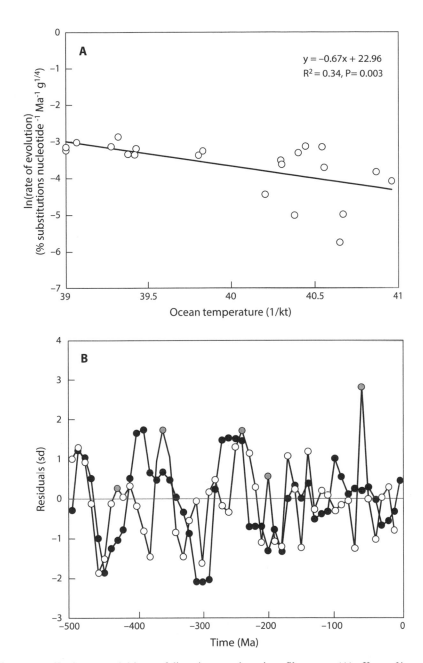

FIGURE 3.1. Environmental drivers of diversity over deep time. Shown are (A) effects of inverse ocean temperature on the rate of evolution in foraminifera; (B) corresponding patterns of temperature variability (black circles) and marine invertebrate species origination rates (white circles) over the last 500 Ma. Gray circles denote the five major mass extinctions. The residuals plotted are mean-standardized after detrending. Redrawn after data in Allen et al. (2006); Mayhew et al. (2012).

270 Ma ago, and persists into the present for many well-studied groups (Stehli et al. 1969). The authors hypothesized that there might be thermodynamic effects on reaction and mutation rates, or, alternatively, a direct effect of incoming solar energy on evolutionary rates. Further supporting this assertion was the observation that the age of coral genera was negatively correlated with water temperature in both the Pacific and the Atlantic Ocean, such that genera that recently evolved were most abundant in the warmest regions of these two oceans (Stehli and Wells 1971). Later, detailed analyses of the fossil record across marine and terrestrial taxa yielded contrasting results and much discussion regarding the effects of temperature on diversity (Mayhew et al. 2007). This controversy was resolved by applying new methods that correct for temporal sampling bias in the global fossil record (Mayhew et al. 2012). The corrected data show a clear positive association between temperature and the rate of evolution over time, and also a positive relationship between temperature and origination rates for marine taxa (see fig. 3.1B). Likewise, on land a latitudinal difference in the rates of molecular evolution (faster in the tropics) has been described in a range of organisms using sister species comparisons of plants, frogs, and mammals (Dowle et al. 2013).

Note, however, that such predictable relationships are not always found in endotherms, since their internal temperature and metabolic rate is largely decoupled from external temperature. Indeed, the rate of molecular evolution of land birds (Bromham and Cardillo 2003) did not show a latitudinal gradient. A comprehensive study of sister taxa of land mammals, however, gave strong evidence of microevolutionary rates being elevated both at low latitudes and at low altitudes, indicating some effect (direct or indirect) of environmental temperature on the speed of evolution (Gillman et al. 2009). Results could not be attributed to other factors thought to influence rates of microevolution, such as body mass or genetic drift. Instead, the results indicated that the tempo of microevolution among mammals is either responding directly to the thermal environment or indirectly via a mechanism such as the "Red Queen" effect of increased evolution driven by rapidly evolving disease agents (Gillman et al. 2009). Similar results were obtained when contrasting low-altitude and high-altitude hummingbird taxa that exist along a smaller-scale temperature gradient (Bleiweiss 1998). In summary, it appears that the evidence for temperature effects on evolutionary rates is strong, but not universal, especially when considering both ectothermic and endothermic organisms.

PRODUCTIVITY

All else being equal, greater energy availability will lead to greater production of organic matter over time (g C m^{-2} a^{-1}), and support a greater density of individuals in both plants and animals that feed on them. This *productivity hypothesis* can act through multiple mechanisms that link increased abundance to increased diversity—for example, through the aforementioned more-individuals hypothesis (Wright 1983)—with the net result that more species coexist in more productive areas. This idea was originally proposed by Hutchinson (1959) and further developed by others (Connell and Orias 1964; Huston 1979; Wright 1983; Rosenzweig and Abramsky 1993). It is related to the classic idea of island biogeography that larger areas support larger populations, which in turn reduces the extinction rate of such populations, leading to higher equilibrium richness, all else being equal (MacArthur and Wilson 1967). Wright (1983) argued that larger populations could also be supported by higher productivity per unit area, mirroring the effect of increased area. Supporting this assertion, habitats of high local productivity, such as rainforests and coral reefs, can indeed show both high individual density and diversity of associated species, and the theory can explain 70–80% of the variation in bird and angiosperm richness on islands worldwide (Wright 1983). Some high-productivity habitats, however, such as marine upwelling areas, often show high density of individuals in combination with low species richness. In this particular case, it would appear there is a better fit with ambient temperature, which is high in speciose coral reefs, but low in species-poor upwelling areas, but this is only one out of many possibilities to explain such patterns. Several authors have attempted to reconcile such conflicting observations by assuming a nonlinear unimodal relationship between productivity and diversity; this is based on theory and experiments that suggest highly productive species outcompete others at the upper end of a productivity gradient (Grime 1973; Huston 1979; Kassen et al. 2000; Kondoh 2001; Worm et al. 2002). Most of these results have been tested on a small, local scale, however, and when richness of plants or animals in habitats of different productivity regimes is analyzed, it is rare that a general relationship is found (Mittelbach et al. 2001; Adler et al. 2011). This may well point to the importance of other confounding factors, which can be resolved only by multivariate analyses. It appears that the effects of productivity by themselves may have limited predictive power, but maybe they need to be considered together with the effects of temperature and area to explain observed gradients in species richness (Jetz and Fine 2012; Valentine and Jablonski 2015). Fortunately, we now have excellent global data sets for these environmental predictors both in the ocean and on land (Field et al. 1998; Hansen et al. 2010), such that their effects can be disentangled. We will examine the relative empirical support for these and other environmental predictors in section 3.4, later.

Environmental Stability and Evolutionary Time

A hypothesis unrelated to energy that considers factors promoting diversity is the *stability-time hypothesis*, which states that environments that have been stable over long periods of time may evolve greater diversity than those that have not

(Sanders 1968). Strictly speaking, Sanders assumed two mechanisms—specifically, lower extinction in more stable environments, and larger numbers of species accumulating over longer time horizons (see table 3.1). In the following, we examine these possible mechanisms.

In contrast to productivity and temperature, the effects of environmental stability and evolutionary time are more challenging to quantify, in part because stability is a multifaceted concept that has no absolute scale (Grimm and Wissel 1997). The stability-time hypothesis originally was proposed to explain the surprisingly high richness seen in some deep-sea communities (Sanders 1968). Briefly, it assumed that old, unperturbed environments like the deep sea would tend to accumulate more species because competitive interactions over time lead to an increasing specialization and narrower niches across the community (Sanders 1968; Grassle and Sanders 1973). Experimental work looking at the effects of environmental stability and time, however, has often come to opposite results, demonstrating a loss of diversity in undisturbed environments over time due to competitive exclusion (Sousa 1979; Paine 1984; Sommer and Worm 2002). Others have pointed out that the original data supporting the stability-time hypothesis were flawed, and observed richness gradients from estuaries to the deep sea can be more easily explained by the difference in total area between these environments (Abele and Walters 1979). As such, stability and time may not generally need to be invoked. On the other hand, it would be hard to argue that evolutionary time has no effect, especially on a macroscale, such as seen for entire ocean basins. For example, the much younger geological age of the Atlantic may well contribute to its lower diversity across taxa, compared to the much older Pacific (Stehli and Wells 1971). Furthermore, global diversity is clearly increasing over time across the entire geological record; thus, there is clearly a temporal dimension that leads to high diversity. Yet the geological record is also punctuated with mass extinction events that set diversity back, on regional or even global scales (Raup and Sepkoski 1982; Alroy et al. 2008). Such environmental perturbations will be the focus of the next section.

3.1.2. Drivers Limiting Diversity

A second class of hypotheses emphasizes the factors that may limit the proliferation of species in challenging environmental regimes—for example, at high latitudes or high altitudes. These include the effects of chronic, seasonal, or episodic environmental stress or disturbance (see table 3.1). Environmental stress is defined as "an action, agent, or condition that impairs the structure or function of a biological system" (Cairns Jr. 2013). Typically, this relates to harsh environmental conditions that can cause an adverse physiological or ecological response. Examples would be changes in temperature, oxygen concentrations, or pH that come close to the tolerance limits of the individual, population, or community in question. While stress can be chronic and ongoing, disturbances are defined as "any discrete event in time that disrupts population, community, or ecosystem structure" (Pickett and White 1985)—for example, in the wake of a fire, flood, or storm. The intensity of stress and disturbances can vary geographically, and over time—for example, through changes in storm frequency over the seasons, or recurring ice ages over millennia—and this can leave an imprint on community composition and diversity. Likewise, pronounced seasonality at high latitudes may limit diversity, as it requires adaptation to a wider range of potentially stressful environmental conditions.

ENVIRONMENTAL STRESS

With respect to environmental stress, there is little doubt that stressful conditions limit the survival of organisms not adapted to these conditions, and hence may impact both community abundance and diversity (Thiery 1982). Few environmental stressors, however, vary consistently by latitude, and dominant stressors vary between land and ocean. Desiccation or water stress, for example, is a major issue on land, but not in the oceans, except for some intertidal environments (Baker 1910). Oxygen depletion, by contrast, strongly affects the distribution and diversity of some marine taxa (Brill 1994; Worm et al. 2005), but is not an issue in most terrestrial habitats. As such, most hypotheses that invoke environmental stress are specific to a particular species group and do not easily scale across taxa and habitats. One universal factor that scales with latitude in the oceans and on land, however, is temperature. Several published hypotheses on the geographical patterns in species diversity emphasize temperature stress (Rohde 1992; Beaugrand et al. 2013; Beaugrand 2014), and the fact that species are limited in their distribution by their widely varying tolerance to suboptimal temperature. These tolerances represent the lower, upper, and optimum temperatures at which species can metabolize, grow, and reproduce. Such physiological limits tend to be fairly stable, a phenomenon that has been described as *niche conservatism* (Wiens and Graham 2005). Experimental evidence for thermal niche conservatism comes from a strain of the bacterium *Escherichia coli* that was maintained over 2000 generations (Bennett and Lenski 1993). Selection for low, medium, and high temperature produced changes in the physiological temperature optimum, but no change at all in the upper and lower limit of the temperature niche. This suggests that within their temperature niches, at least some organisms are able to adapt to fluctuations or directed changes in temperature, but that this does not affect their absolute thermal niche. However, the same authors observed slight (1–2 degree) changes in niche width when selection pressures operated at the lower thermal range boundary (Mongold et al. 1996). Likewise, the paleontological record also suggests that clades that originated in the tropics can evolve tolerance to cooler temperatures; however, substantial time lags (~5 Ma) between the origins of tropical clades and their expansion into the temperate zone suggest that this process occurs rather slowly, if at all (Jablonski et al. 2013).

From this arises a question about the physiological mechanisms that underlie the observed conservatism of thermal niches. Or as Hutchinson (1959) famously asked: If one species can adapt to colder environments, why don't they all do it? (Note that this question applies to other stressors as well.) In a book devoted to the topic, Johnston and Bennett (1996) summarize the evidence for temperature adaptation in animals from a physiological and evolutionary perspective. Adaptation to low temperature emerges as a multifaceted problem, involving a large number of specialized mechanisms at the molecular (for example, changes to key proteins and enzymes), cellular (changes to cytoskeleton and membranes), and physiological (changes to circulatory system) levels. This may suggest that the basic metabolic "machinery" of life evolved at warm temperatures and that colonization of cold environments (that is, high latitudes, high altitudes, and the deep sea) involves a suite of "evolutionary fixes" that make life at low temperature possible. There also seem to be strong trade-offs involved, resulting in narrower temperature niches in polar species than in warm-adapted ones. An example are the Antarctic cod icefishes (Nothotheniidae), which feature many specialized adaptations to survive subzero temperatures, but cannot survive at ambient temperatures greater than 6 to 8°C.

The absolute width of thermal niches may also be related to the variability experienced in a given environment. This is *Rapoport's rule*, which states that species at higher latitudes should have larger latitudinal ranges (Taylor and Gaines 1999), because seasonal temperature variability is larger at higher latitudes, and thus species need to be tolerant of a range of conditions, or perform seasonal migrations, both of which could result in larger ranges. This is particularly true on land, where absolute fluctuations are much higher than in the sea, particularly near the poles.

Given these findings about the importance of temperature niches, it may not be surprising that the latitudinal range limits of many species can be predicted well from their temperature tolerances (Boyce et al. 2008; Sunday et al. 2012), and observed temperature changes tend to go in lockstep with changing species ranges at seasonal (Whitehead et al. 2008) as well as interannual scales (Pinsky et al. 2013). There appear to be some interesting contrasts between terrestrial and marine species, however: Sunday et al. (2011) found that upper and lower thermal limits both decreased with increasing latitude in marine species—that is, total niche breadth was constant, on average. On land, however, the upper limit remained similar, while lower limits decreased, possibly related to higher seasonal variability in high-latitude terrestrial environments. They also showed that marine organisms tended to occupy their thermal niches more fully, such that their fundamental niche equaled the realized niche. As a result, both upper and lower latitudinal boundaries were found to shift predictably with climate change in marine species (Sunday et al. 2012). On land, in contrast, the warm end of the range was often not filled, and hence the lower limit of the latitudinal range did not shift as predictably, often resulting in a time lag in the response of the trailing range edge to climate warming (Sunday et al. 2012).

These observations suggest strong thermal control on species ranges in recent times, complementing similar findings from the past. For example, warming during the last interglacial, about 125,000 years ago, appears to have displaced the ranges of reef-building coral species to higher latitudes (Kiessling et al. 2012). Hence, the spatial association of surface temperature, species ranges, and, by inference, diversity patterns seems to hold over time and space, though potentially varies in its strength (for example, Yasuhara et al. 2012).

Some models attempt to explain patterns of diversity exclusively based on observed present-day temperature tolerances—this may be termed the *thermal niche hypothesis* (Beaugrand et al. 2013). When assembling a wide range of randomly generated stenothermic and eurythermic "pseudospecies" along a global gradient, and accounting for insufficiently filled niches toward the poles (thought to result from repeated glaciations), Beaugrand and coworkers (2013) reconstructed the latitudinal diversity gradient for foraminifera and copepods with >80% accuracy. Note, however that the chosen model represented one of 79 tested configurations that were used to fit to the data. Also, both of these groups are pelagic and planktonic, and differ markedly in their latitudinal pattern from known coastal, terrestrial, and deep-sea species groups (see chapter 2), suggesting that thermal niches may not offer a general explanation for observed diversity patterns. Nevertheless, such results suggest that thermal niches can help us understand some of the factors that limit species distributions, and hence diversity, as reviewed extensively by Beaugrand (2014).

DISTURBANCE

Disturbances are, by definition, discrete and episodic events that disrupt communities and influence diversity in many ways. Large disturbances, such as ice ages, could have shaped geographic patterns of diversity by driving up extinction rates at high latitudes and resetting the evolutionary trajectory (Fischer 1960). There is good evidence that the ice ages have had a lasting effect on communities at higher latitudes, particularly on land (Comes and Kadereit 1998). However, it is not clear that the legacy is one of reduced species richness; indeed, these disturbances can lead to reproductive isolation of populations, genetic divergence, and speciation (Hewitt 1996). Moreover, few studies have found general support for elevated extinction rates at high latitudes, and there is a general sense that differences in extinction rates alone cannot sufficiently explain observed geographic gradients in species richness (Hawkins et al. 2006; Jablonski et al. 2006; Dowle et al. 2013).

Only marginally related to the question of global gradients in biodiversity, the *intermediate disturbance hypothesis* does not consider speciation and extinction as much as ecological interactions within a community. It predicts high diversity under "intermediate" rates of disturbance, where competitive exclusion by dominant species is prevented. While Connell (1978) used this hypothesis to explain high richness in coral reefs and rainforests, later tests of the hypothesis found that it has reasonable predictive power in rocky shore and plankton communities, albeit at a local scale (Sousa 1979; Petraitis et al. 1989; Floeder and Sommer 2000). Like many hypotheses that focus on species interactions, the intermediate disturbance hypothesis is more concerned with mechanisms that explain the maintenance of species diversity in a particular community, rather than the generation of diversity gradients at global scales on land or in the sea.

SEASONALITY

One of the obvious differences between tropical and temperate environments is the pronounced seasonality found in the latter, which necessities specialized adaptations such as hibernation, migration, and larger thermal niches. Larger variation in environmental conditions may cause mortality and possibly local extinction— for example, during harsh winters—thus limiting diversity (*seasonality hypothesis*; Fischer 1960). Another aspect is food supply, which is much more seasonal and episodic at high latitudes, requiring a shift from specialist to generalist feeding strategies. Taylor and Taylor (1977) observed such a shift in feeding patterns among predatory gastropods in the North Atlantic, concomitant with a large decrease in diversity. This decrease was most marked around 40 degrees North latitude, a point at which primary production changes from a less seasonal to a strongly seasonal regime (Taylor and Taylor 1977). Later authors, however, did not find a consistent correlation with seasonality across species groups (Currie 1991), and a similar breakpoint in diversity around 40 degrees latitude is not generally seen in the maps in chapter 2. Like environmental stress and disturbance, it appears that seasonality may explain aspects of individual patterns in diversity, but might not present the level of generality across taxa that is evident in the global patterns documented in chapter 2.

3.1.3. The Effects of Habitat Area and Quality

Finally, it is intuitive that the size and quality of the available habitat area may play a role in setting at least an upper limit on the number of species that can coexist there (Connor and McCoy 1979). Larger areas can support larger populations, all else being equal, similarly to areas of higher productivity, potentially leading to increased diversity through mechanisms such as reduced extinction rates or sampling from the regional species pool (the more-individuals hypothesis). Larger areas also often feature a greater diversity of habitat types, potentially providing a greater variety of niches for species to inhabit. Finally, the physical complexity and shape of a habitat (for example, a range of elevations, or structural complexity of a forest versus a grassland), and the trophic complexity of associated food webs, may further elevate observed richness of species occupying those habitats.

HABITAT AREA

The simple observation that larger areas typically yield greater species counts is formalized in the *species-area relationship* (SAR)—without doubt one of the most general ecological laws (Preston 1962; MacArthur and Wilson 1967; Connor and McCoy 1979; Hubbell 2001; Drakare et al. 2006). The SAR captures the relationship between the area of a habitat (A), and the number of species (S) found within that habitat. Empirically, the relationship is most often fitted by a power function of the form

$$S = cA^z.$$

Here, z denotes the linear slope of the relationship in log-log space, and c the intercept. A meta-analysis of 794 SARs confirmed the relationship's generality, but also found that its parameters can vary systematically with latitude and body size (steeper slopes in the tropics and for large organisms), habitat, as well as sampling scale and design (Drakare et al. 2006).

How can this pattern be explained mechanistically? MacArthur and Wilson explored the SAR from first principles in their theory of island biogeography (MacArthur and Wilson 1967), modeling the balance of immigration and extinction on islands of different size and isolation. Larger islands support larger populations on average, which are predicted to result in lower extinction rates, and hence more species will coexist on a larger island, all else being equal. Building on MacArthur and Wilson's work, the neutral theory of biodiversity and biogeography generalizes to communities not located on islands by constructing a neutral metacommunity with constant per capita rates of dispersal (immigration and emigration), speciation, and extinction (Hubbell 2001). The model also yields a species-area relationship that closely follows empirically observed patterns. Hence, there is both a solid theoretical as well as an empirical foundation for the effects of increasing habitat area on species richness. Note that the SAR and the productivity hypothesis share at least one common mechanism—specifically, that both larger areas and more productive ones will support larger populations that are less likely to go locally extinct (see table 3.1). Hence, Wright (1983) treated the effects of productivity and area on species richness as additive. In his empirical analysis, the best predictor for bird species richness on islands was total island area multiplied by the rate of primary productivity per unit area. More recent work on birds, however has suggested more complex scaling relationships, in which both area and productivity scale with richness, but to different degrees in different regions (Hurlburt and Jetz 2010).

Habitat Complexity

Another aspect of the environment that may affect diversity is the number of different habitat types, which also tends to increase with sampling area, and correlates well with species richness. Both effects (larger populations and more habitat types) may contribute to the observed rise in species richness with habitat area described by the SAR, but these mechanisms are not explicitly captured by the theory of island biogeography or the neutral theory of biodiversity. Their relative importance can be analyzed by quantifying the effects of surface area, the number of habitats types (for example, different vegetation types), and the structural complexity of the habitats (for example, elevational range). Some aspects of habitat complexity are physical (for example, topographic complexity) while others are biological (for example, structural complexity of biogenic habitats such as forests and reefs). There is a clear sense that more complex habitats allow for a greater range of species to coexist, owing to a larger variety of environmental conditions and food sources. Likewise, trophic complexity may beget greater diversity by allowing for more niches, such as specialized predators, parasites, and the like. Some aspects of trophic complexity may also act to maintain high diversity, such as the action of a keystone predator maintaining diversity of prey items by preventing competitive exclusion; this is the *predation hypothesis* (Paine 1966). These two aspects, increasing specialization and trophic complexity, point at potential mechanisms by which "diversity may beget diversity" (Brown 2014). Such mechanisms help to explain the extraordinary diversity found in some complex habitats, but cannot explain the emergence of a geographic gradient in the first place, as they do not identify the causal mechanisms behind initial diversification (Rohde 1992).

HABITAT DOMAIN

The spatial domain within which habitat and species assemble geographically may influence the spatial pattern of diversity. A much-discussed example is the *mid-domain hypothesis*. This is different from the previously discussed theories in that no environmental driver is assumed to operate here, but purely an interaction of geometry and species ranges. The mid-domain hypothesis is a null-model that assumes random placement of species ranges across a habitat area, or *domain*. This will result in maximum richness near the center of that spatial domain, such as a continent or an ocean basin. The argument is that the observed maximum richness in the central Indo-Pacific, for example, could be predicted by such a null-model (Colwell and Lees 2000). Empirical support for this hypothesis has been quite weak at biogeographic scales, however (Hawkins and Diniz-Filho 2002; Zapata et al. 2005; Currie and Kerr 2008), indicative of nonrandom placement of species ranges. Yet this hypothesis has invigorated an interest in null-models in ecology to be confronted with available data. Recently, Beaugrand et al. (2013) extended the null-expectation of a spatial mid-domain effect to niche space, assuming random placement of temperature niches within the bounded range of observed sea surface temperatures in the global ocean. As discussed earlier, this randomly generated pattern fit reasonably well to the observed richness gradients in pelagic zooplankton, but required some additional modifications due to the assumed effects of ice ages reducing richness at high latitudes (Beaugrand et al. 2013). We will return to null-models of species richness in chapter 4.

3.2. SPATIAL AND TEMPORAL SCALE

When considering the drivers and environmental predictors of biodiversity, one needs to think carefully about processes operating across multiple spatial and temporal scales. For example, species richness patterns are known to correlate with different predictors at different spatial scales (Jetz and Rahbek 2002; Belmaker and Jetz 2011; and fig. 3.2). Notably, net primary productivity tends to be a dominant environmental predictor at smaller scales, possibly by increasing the number of individuals that can coexist. At larger scales ($\geq$1000 km), however, temperature was found to be the strongest predictor of species richness in amphibians, birds, and mammals on land (fig. 3.2). At the same time, the predictive power of these correlates, and the total variance that was explained, increased sharply at larger scales, likely because the pattern is more reflective of larger-scale processes, and less sensitive to local conditions, and environmental noise.

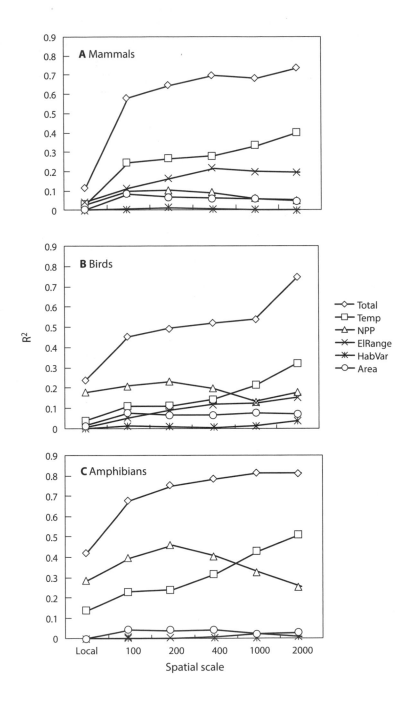

Such results reflect an important reason why we are focusing our discussion on large spatial scales and the general ecological and evolutionary mechanisms operating at those scales. As discussed in the previous chapter, this is also why we map all data to the same coarse grid, avoiding the scale dependence and reduced explanatory power mentioned earlier, and also minimizing uncertainty and spurious overprecision associated with some spatial data. Local diversity patterns are likely shaped by an alternative set of mechanisms (see fig. 3.2), including but not limited to local species interactions, environmental constraints, and even human impacts on local habitat structure and biodiversity (Sommer and Worm 2002). These factors may determine which species from the regional species pool will be able to colonize, reproduce, and maintain a viable population at a local site. While these are clearly important questions—for example, in the context of local-scale management in the face of environmental change—we are primarily concerned with the factors that have determined the size and distribution of the regional species pool in the first place.

We also have to be aware of different processes operating at different temporal scales. First, there is the geological timescale (10–100 Ma). At this scale, major tectonic events arc changing the face of the planet, affecting the size and geography of ocean basins and continental margins. The fossil record suggests that hotspots of marine biodiversity wax and wane in the wake of such major tectonic events: they persist for some time (tens of Ma), and then fade as new geological events reshape these habitats (Bellwood et al. 2012). These are the previously discussed "hopping hotspots" (Renema et al. 2008), which provide a dynamic deep-time perspective on the spatial organization of marine biodiversity (see fig. 2.10). Second, there is the evolutionary timescale (0.01–10 Ma). At this scale, macroevolutionary processes play out within a particular geological setting. For example, repeated changes in sea level can bring about isolation of islands, atolls, and semi-enclosed seas, thereby creating spatial barriers that lead to speciation (the so-called species pump; Bellwood et al. 2012). Third, at the ecological timescale (<0.01 Ma), shorter term climatic and biological processes may prevail,

FIGURE 3.2. Analyzing environmental predictors of vertebrate species richness across scales. Spatial scale represents the approximate diameter (km) of the spatial units that were used to analyze global relationships between hypothesized environmental correlates and species richness. While the effects of annual net primary productivity (NPP) dominate at smaller scales, temperature (Temp) becomes the dominant environmental predictor at larger scales. Total predictive power and variance explained (Total R^2) also increased at larger spatial scale. Elevation range (ElRange), habitat variety (HabVar), and area (Area) are less important, particularly at the larger spatial scales. Redrawn after data in Belmaker and Jetz (2011).

which play out in response to recent environmental and biotic conditions—for example, conditions encountered during the current Holocene interglacial. At this scale, dispersal, competition, succession, and other ecological processes likely influence large-scale biodiversity patterns. Finally, the present time (<0.001 Ma) is increasingly shaped by human forces, and has been deemed to represent its own geological epoch, the Anthropocene (Waters et al. 2016). At this moment in geological history, anthropogenic drivers are becoming ever more important, including the effects of exploitation, land use and climate change, ocean acidification and habitat destruction (Palumbi 2001; Halpern et al. 2008). These processes are already reshaping marine and terrestrial biodiversity patterns (Pimm et al. 1995; Pinsky et al. 2013; Dirzo et al. 2014; McCauley et al. 2015).

It is important to note that an empirical analysis of present-day patterns and their environmental correlates will primarily shed light on processes operating at recent timescales, operating under currently realized environmental conditions. Where possible, and where appropriate evidence exists, we will attempt to also discuss changes in the magnitude of major drivers, and their effects on shaping species richness patterns over time. Through our model development in chapter 4, we will then aim to integrate processes that operate at both evolutionary and ecological timescales.

3.3. EMPIRICAL PREDICTORS OF DIVERSITY

Typically, the relative importance of hypothesized drivers of diversity is evaluated via statistical analyses that examine covariation in the spatial patterns of diversity and various present-day environmental predictors, such as temperature, productivity, habitat area, among others (Gaston 2000; Kreft and Jetz 2007; Tittensor et al. 2010; Woolley et al. 2016). While such analyses can not prove causation, and can be complicated by statistical problems such as colinearity and spatial autocorrelation (Dormann et al. 2007), they nevertheless provide important empirical grounding. Tables 3.2 to 3.5 and fig. 3.3 present published evidence identifying primary and secondary environmental predictors for the species richness of well-studied taxonomic groups with global sampling coverage. Here, we discuss these organized by the major habitats that we originally described in chapter 2. We consider only those studies that tested and compared multiple hypotheses, and that used suitable analysis methods that corrected for spatial autocorrelation such as spatial linear models or simultaneous autoregressive models. For species groups where several publications were available, we selected the one with the most comprehensive data set, to avoid pseudo-replication of studies using different versions of the same data.

TABLE 3.2. Environmental Predictors of Species Richness in the Coastal Realm

Variable	Fishes	Sharks	Pinnipeds	Bivalves	Corals	Snails	Cephalopods	Seagrasses	Mangroves
SST	10.7***	7.1***	−10.0***	7.3***	7.7***	8.6***	7.1***	4.4***	9.3***
SST slope		2.4*	4.3**						
Coastline length	7.9***	13.0***	4.5***	5.7***	3.1**	4.1*	6.5***	4.3**	2.0*
Primary productivity		3.6**	5.5***			−2.2**			
SST range		−2.5*	−3.2**			3.7**			
Oxygen stress									
Indian Ocean	3.7**				3.8**	7.0***	−1.8*	2.6*	
Pacific Ocean	4.5***				3.5**	6.9***	−2.8**	2.0*	
Pseudo R^2	0.71	0.75	0.88	0.97	0.73	0.78	0.89	0.75	0.85

Note: Shown are results of minimal-adequate spatial linear model for environmental correlates (SST = sea surface temperature). Numbers are z-values; asterisks represent significance levels at $P < 0.05$ (*), 0.01 (**), or 0.00001 (***). Ocean column z-values represent contrast against Atlantic Ocean. Species groups and scale of analysis correspond to patterns shown in figs. 2.2 and 2.3, with the exception of ophiuroids, which can be found in table 3.2. Individual drivers are shown in fig. 3.4.

Source: After data in Tittensor et al. (2010), Valentine and Jablonski 2015, and IUCN 2016.

TABLE 3.3. Environmental Predictors of Species Richness in the Pelagic Realm

Variable	Tuna/billfish	Sharks	Seabirds	Cetaceans	Foraminifera	Euphausiids	Squids
SST	7.0***	11.8***	31.4***	6.6***	16.6***	6.6***	4.0**
SST slope	3.1**	—	—	—	3.3**	3.9**	2.7**
Coastline length	—	5.8***	—	—	-2.8**	—	—
Primary productivity	—	—	54.5***	12.1***	—	3.4**	—
SST range	-3.6**	—	—	—	—	-7.8***	—
Oxygen stress	—	—	—	—	-2.3*	—	—
Pseudo R^2	0.76	0.81	0.67	0.89	0.79	0.85	0.88

Note: Shown are results of minimal-adequate spatial linear model for environmental correlates. Numbers are z-values; asterisks represent significance levels at $P < 0.05$ (*), 0.01 (**), or 0.00001 (***). Ocean column z-values represent contrast against Atlantic Ocean. Species groups and scale of analysis correspond to patterns shown in figs. 2.4 and 2.5. For seabirds only; note that additional effects for wind energy and distance to the coast were also described for that taxon.
Source: After data in Tittensor et al. (2010) and Davies et al. (2010).

TABLE 3.4. Changes in Environmental Predictors with Depth

Variable	Depth and habitat		
	20–200 m	200–2000 m	2000–6500 m
	Shelf	Slope	Deep sea
Annual mean temperature	11.49***	3.61***	—
(Annual mean temperature)^2	—	−2.71**	—
Annual mean oxygen	−2.17*	—	—
Seasonal variation in NPP	3.54**	—	1.61**
(Seasonal variation in NPP)^2	−2.48*	—	—
Particulate organic carbon flux	−4.43*	−3.06*	3.09**
(Particulate organic carbon flux)^2	2.13*	—	−2.46*
Distance to continental margin	—	—	0.45*
(Distance to continental margin)^2	—	—	—
Oxygen stress (OMZ)^2	1.71*	—	—
Pseudo-R^2	0.35	0.37	0.21

Note: Shown are results from spatial linear models for brittle star (Ophiuroidea) species richness across three depth habitats. Both linear and quadratic effects (in brackets) were included. Only significant terms (z-values) are given; asterisks represent significance levels at $P < 0.05$ (*), 0.01 (**), or 0.001 (***). Distance to continental margin is applicable only for deep-sea species.

Source: After data in Woolley et al. (2016).

TABLE 3.5. Environmental Predictors of Vascular Plant Species Richness on Land

Variable	Coefficient	z-value
Area	0.118	11.5***
PET	0.747	12.4***
Wet days	0.542	12.3***
Habitat complexity	0.01	11.3***
Vegetation structure	0.022	4.5***
Floristic kingdom		
Nearctic	0.081	1.7
Australis	0.162	2.2*
Capensis	0.281	4.1***
Paleotropic	0.062	1.5
Palaearctic	0.023	0.5
Pseudo R^2	0.702	—

Note: Shown are results from spatial linear models across 1032 geographic regions worldwide. Asterisks represent significance levels at $P < 0.05$ (*) or 0.001 (***). PET (potential evapotranspiration) is a combined measure of temperature, solar insolation, and day length. Habitat complexity is a compound variable that captured altitudinal range and number of habitat types. Vegetation structure describes the three-dimensional complexity of vegetation.

Source: After data in Kreft and Jetz (2007).

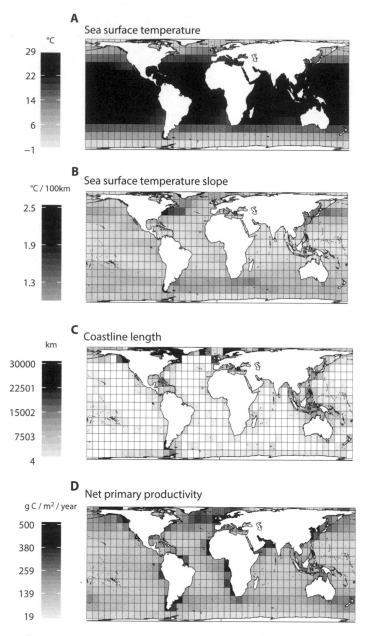

FIGURE 3.3. Present-day environmental predictors of species richness in: the surface ocean (A–D), the deep-ocean (E–F) and on land (G–I). These predictors empirically represent the best correlates of species richness overall, see also tables 3.2 to 3.4 and chapter 5. A–D after data in Tittensor et al. (2010); E–F after data in Woolley et al. (2016); G–I after data in Kreft & Jetz (2007) and Climatic Research Unit Global Climate Dataset (http://www.ipcc-data.org/observ /clim/cru_climatologies.html).

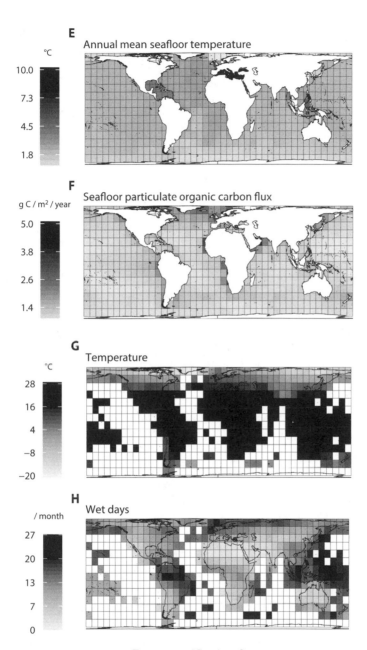

FIGURE 3.3. (*Continued*)

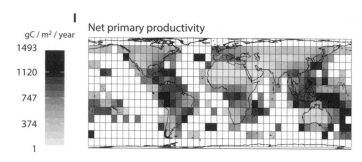

FIGURE 3.3. (*Continued*)

3.3.1. Empirical Predictors of Species Richness in the Oceans

Environmental predictors that are commonly analyzed in the oceans are displayed in fig. 3.4 at the same resolution as the species richness data in chapter 2. These relate to the hypotheses mentioned earlier as follows: Sea surface temperature (SST) and net primary productivity (NPP) relate to the effects of thermal energy (evolutionary-speed hypothesis and thermal-niche hypothesis) and productivity (more-individuals hypothesis). Productivity in the deep sea, however, is not directly fueled by NPP, but by export production raining down from surface waters, and measured as particulate organic carbon (POC) flux (Tittensor et al. 2011; McClain et al. 2012; Woolley et al. 2016), thus replacing NPP as a predictor of species richness (more-inidviduals hypothesis). Moreover, a major stress factor unique to the ocean is low oxygen concentration (<2 mL L^{-1}), found regularly in upwelling areas, and expanding elsewhere (stress hypothesis; Stramma et al. 2010). The length of coastline (km) relates to the area of coastal habitat (habitat-area hypothesis; Tittensor et al. 2010). The steepness of the SST gradient approximates the availability of thermal fronts, a critical habitat feature in pelagic waters (habitat-area and habitat complexity hypotheses; Olson et al. 1994). The standard deviation in SST and seasonal variability in NPP can both be used to test the seasonality hypothesis.

When analyzing statistical relationships between these predictors and richness of coastal and pelagic species groups, the spatial pattern of species richness was most strongly and consistently related to sea surface temperature (see tables 3.2 and 3.3), and the univariate relationship between temperature and species richness was either near-linear (in most coastal taxa) or saturating (in most pelagic taxa; see fig. 3.4). This finding corresponds to patterns of coastal species richness peaking equatorially or in the tropics (see figs. 2.1 to 2.4) and pelagic species peaking in the subtropics (see figs. 2.5 and 2.6), respectively.

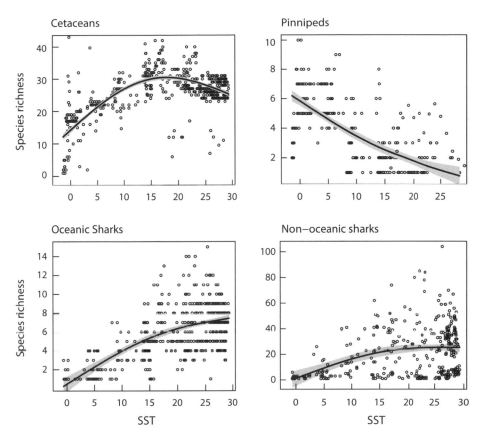

FIGURE 3.4. Univariate relationship of species richness and sea surface temperature in coastal and pelagic taxa. Note distinct relationships for endotherms (pinnipeds, cetaceans, and seabirds). Each point is a cell in the global equal-area grid. Trends indicated by smoothed fit from generalized additive model with basis dimension 3. Gray shading indicates 95% confidence limits. Updated and expanded; after data in Tittensor et al. (2010), Davies et al. (2010), and Valentine and Jablonski (2015).

Of secondary importance, but still generally influential, were the effects of habitat area and complexity. The length of coastline was an important and consistent predictor of species richness in many coastal taxa (see table 3.2), with more habitat area allowing for greater species richness (note that longer coastlines often also entail greater complexity such as found in the world's large archipelagos, specifically the Indonesian-Australian Archipelago and the Caribbean). Notably, the Arctic Archipelago, despite its massive coastline, supports few species, perhaps indicating the primacy of the effects of temperature over habitat area and complexity.

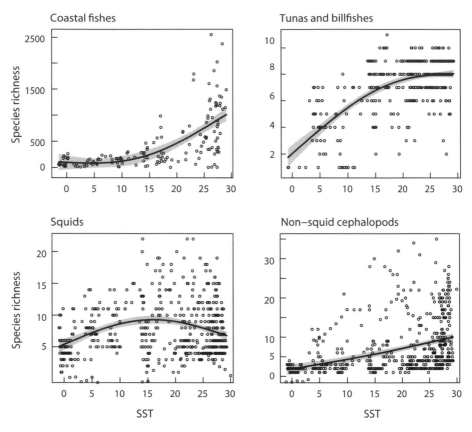

FIGURE 3.4. (*Continued*)

Likewise, in pelagic taxa, the second most influential predictor of diversity was sea surface temperature (SST) slope, which indicates the steepness of oceanographic fronts and transition zones that are commonly used by pelagic organisms as feeding habitats and migratory routes (Haney 1986; Olson et al. 1994; Polovina et al. 2001; Ferraroli et al. 2004). Steep frontal areas add habitat complexity, commingle individuals through advection, and aggregate species across diverse taxa. As such, these pelagic habitat features may act similarly to the more solid coastline habitats for near-shore organisms, or topographic complexity for terrestrial taxa.

It is interesting that the primacy of temperature as a positive predictor of diversity was not seen as clearly for endothermic groups such as pinnipeds, cetaceans, and seabirds (see tables 3.2 and 3.3). These endothermic species groups may be partly decoupled from the constraining effects of ambient temperature in terms of their distribution but also their evolutionary speed: no significant difference

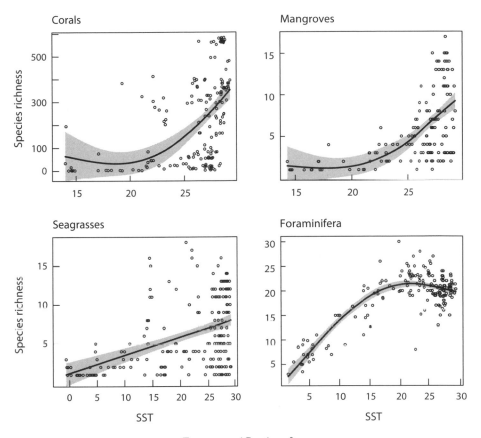

FIGURE 3.4. (*Continued*)

was seen in diversification rates of birds across a global latitudinal gradient (Jetz et al. 2012b). Yet this partial independence from ambient temperature comes at a cost of high metabolic demands, which require high ambient productivity to satisfy (Lavigne et al. 1986). Hence, many endothermic species may converge around areas of high productivity, irrespective of water temperature. Indeed, productivity is a strong predictor for marine mammal species richness, which tends to peak in high-latitude convergence zones or low-latitude upwelling areas, both of which are cool and nutrient-rich (Schipper et al. 2008); similar patterns are seen in seabirds (Davies et al. 2010). Where sufficient productivity is available to support high-energy lifestyles in cooler waters, the enhanced metabolic rate found in endotherms may provide competitive advantages over ectothermic species.

The second group that was not primarily predicted by ambient temperature was deep-sea species (see table 3.4). While the richness of shelf and slope ophiuroids

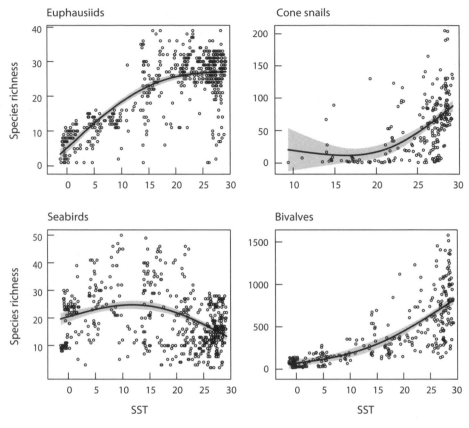

FigURE 3.4. (*Continued*)

was still strongly and positively related to ambient temperature, deep-sea richness was much more strongly related to patterns in export production, and the seasonality in NPP, which drives large pulses of export production during seasonal blooms (see table 3.4; Woolley et al. 2016). Similarly, for gastropods, bivalves, and nematodes analyzed in the Atlantic Ocean, export productivity from surface waters was consistently the best predictor of species richness (Lambshead et al. 2000; Lambshead et al. 2002; Tittensor et al. 2011). This likely relates to the fact that, except for some isolated basins such as the Mediterranean, temperature is near-uniformly cold across the deep sea, typically ranging between 2 and 4°C. Thus, metabolic rates will not vary dramatically in these habitats and are more likely to respond to local changes in food supply than temperature. Hence, while thermal energy (proxied by temperature) clearly emerged as a primary environmental predictor for most groups, productivity was also important for species that experience little temperature variation, either because they are endothermic (near-constant warm body temperature) or live in the deep sea (near-constant cold body temperature).

When considering such statistical associations between present-day environmental variables and diversity, there is of course always the question of causation. A static pattern provides weaker evidence for causal relationships than a dynamic pattern, especially as the modern-day environment may not reflect past influences. Paleo-ecological studies can be useful in this regard, as the association between the environment and diversity can be tracked across changing conditions. Taxa that form abundant microfossils are particularly well-suited for this work, because they can be retrieved from sediment cores across all oceans in large numbers, providing a comprehensive image of global diversity patterns through time (Yasuhara et al. 2015).

Foraminifera, for example, form small exoskeletons that fossilize well and are readily identified to species. As for most other marine groups, the present richness pattern of pelagic foraminifera correlates closely with sea surface temperature, which explained 90% of the observed variation in foraminiferan richness (Rutherford et al 1999). Other variables, such as chlorophyll *a* concentration, salinity, nitrate, or mean solar irradiation, were much less powerful predictors of richness for this taxon. Subsequent analysis of temporal changes in diversity patterns for pelagic foraminifera supported the overriding influence of temperature variation as a driving force (Yasuhara et al. 2012). While the strength of latitudinal gradients adjusted dynamically to changes in average water temperature across glacial and interglacial cycles over the last 3 million years, the diversity-temperature relationship remained remarkably constant (Yasuhara et al. 2012). This suggests a causal relationship that is driving biogeographic responses to variation in climate at long timescales. Interestingly, a similar adjustment is often observed at much shorter timescales, corresponding to seasonal and interannual variation in water temperature (Whitehead et al. 2008; Pinsky et al. 2013), again suggesting a causal relationship that tracks temperature variation over space and time.

Of course, such a relationship may not apply equally to each taxonomic group, or at all spatial scales. While species richness of most coastal invertebrates shows a strong relationship to temperature at a global scale (Tittensor et al. 2010; Valentine and Jablonski 2015), for example, Fernandez et al. (2009) found reverse latitudinal gradients (and relationships to SST) among benthic invertebrates with different life histories along the coast of Chile. Species with planktonic larvae showed a strong positive relationship with temperature, but those with direct development showed the opposite pattern. This effect of life history held true both in crustaceans and mollusks (Fernandez et al. 2009). The authors hypothesized that such a pattern might be explained by the effect of temperature on length of planktonic life (longer in colder climates) and on brooding costs (higher in warmer climates). Direct development may be an advantageous strategy in colder climates; hence, the greater diversification of species with this developmental mode. Thus, species with unusual life histories or adaptations may deviate from the more common pattern of a strong positive relationship to temperature. This is very obvious for cold-water specialists such as pinnipeds, penguins, and Antarctic icefishes, which naturally show their greatest abundance and richness in cold climates, and only secondarily diversify into warmer waters. Thus, the observed richness of these cold-adapted taxa correlates negatively with ambient temperature (see pinnipeds in table 3.2). Note, however, that at higher levels of taxonomic aggregation (classes, phyla, and kingdoms), a generally positive relationship with temperature reemerges. This suggests that while, in this example, mammals, birds, and fishes show peak richness at lower latitudes and warm ambient temperatures, some subgroups (pinnipeds, penguins, and icefishes) specialize in cold (or otherwise extreme) climates, and develop their greatest diversity there.

3.3.2. Empirical Predictors of Species Richness on Land

In contrast to the marine environment, terrestrial habitats are uniquely limited by the availability of ambient water. It took hundreds of millions of years for macroscopic marine life forms to evolve specific adaptations (such as seeds, eggs, and thick cuticulas) that enabled them to colonize the land. Hence, it is not surprising that standing biomass and production, as well as species richness on land relate to the availability of water, as well as the availability of energy.

In a comprehensive geospatial analysis, it was found that environmental factors related to temperature, moisture, habitat area, and complexity predicted plant diversity at a global scale (Kreft and Jetz 2007) (see table 3.5). *Potential evapotranspiration* (PET) gave a better prediction than temperature or net primary productivity. PET is a compound measure of ambient energy input and is calculated from average temperature, total insolation, and day length. The number of wet days emerged as an important secondary predictor at the global scale (see table 3.5). Habitat area and complexity (topographic complexity and habitat types) were also important, and there were some important differences among biogeographic provinces, likely related to historical factors (see table 3.5). Notably, these results do not support a primary role of productivity, which was often assumed to drive species richness on land (for example, Huston 1994). Thus, the richness of plants is primarily driven by environmental factors relating to energy, moisture, and habitat, at least at larger scales.

This finding may apply more broadly to terrestrial taxa. A comprehensive literature review concluded that measures of ambient thermal energy, water, or water-energy balance explained spatial variation in richness better than other climatic and nonclimatic variables in 82 of 85 cases (Hawkins et al. 2003). Even when considered in isolation, water and energy variables explained on average over 60% of the variation in the richness of a wide range of plant and animal groups (Hawkins et al. 2003). As discussed earlier, however, this strong link to climatic variables is most prominent at large scales, particularly for mammals and birds, for which diversity is more likely to relate to local productivity at smaller spatial scales (Belmaker and Jetz 2011). This suggests that temperature is more important in driving regional, but less so local, richness in these taxa (Belmaker and Jetz 2011; see fig. 3.2).

The strong convergence between the richness patterns of vertebrate consumers and plants could be causal as well, such that plant richness is driven by climatic variables, but animal richness relates to the number of plant species (that is, diversity begets more diversity; Brown 2014). While there is good evidence for such a direct effect at small scales (in individual habitats), it has been shown that at large scale vertebrates and plants are both driven by climatic variables, rather than causally related to each other (Jetz et al. 2009).

While much of the terrestrial research has focused on vertebrates and plants, newer studies include less conspicuous taxa such as microorganisms and fungi. A study of soil samples from North and South America concluded that microbial diversity was mostly related to soil pH, with maximum diversity at neutral pH (7 to 8), and declining diversity at higher or lower pH (Fierer and Jackson 2006). Fungi are another poorly known group, despite the fact that they may be one of the most species-rich taxa on land. When analyzing the diversity of soil fungi across all continents, it appeared that species richness was strongly linked to moisture and soil chemistry, particularly the availability of calcium (Tedersoo et al. 2014). In contrast to animal taxa, richness of all fungi and functional groups was unrelated to plant diversity, with the exception of ectomycorrhizal root symbionts. This suggests that plant-soil feedbacks do not influence the diversity of soil fungi at the global scale (Tedersoo et al. 2014).

Freshwater taxa offer another interesting contrast because their habitats are embedded in the terrestrial sphere but are not moisture-limited. Across freshwater vertebrates (mammals, birds, amphibians, and fish) and crayfish, ambient energy (temperature and solar radiation) consistently explained a majority of variation in total species richness and endemism (Tisseuil et al. 2013). Habitat area and historical factors related to dispersal explained much of the remaining variation. The modeled response to changes in ambient energy was stronger than for other factors and near-linear (Tisseuil et al. 2013). This shows that the diversity of freshwater taxa appears to be driven by similar factors as land taxa, with the exception of moisture.

In conclusion, it appears that the diversity of life on land is primarily related to the availability of ambient energy, measured as temperature or PET, in combination with sufficient moisture, habitat area, and complexity. Thus, there is a fundamental similarity between primary correlates of diversity on land and in the sea, considering similar primacy of ambient energy and habitat as leading predictors of species richness. Beyond this, there is the unique dependence on ambient moisture that characterizes life on land, where water is often limiting, and the greater oxygen constraints in the ocean. Of course, these factors have all changed over time—for example, between glacial and interglacial cycles. Analyzing both historic and present-day conditions, it was shown that a combination of current temperature and historically integrated habitat area and productivity explained up to 87% of observed variation in terrestrial vertebrate richness (Jetz and Fine 2012). Similar conclusions were reached for marine species, particularly the well-studied mollusks (Valentine and Jablonski 2015), and for vascular plant richness on islands, which carries a large signal of historical climate and area changes, especially for endemic species (Weigelt et al. 2016).

3.4. SYNTHESIS

Three major classes of hypotheses seek to explain species richness patterns on our planet. These relate to (1) factors that may promote high diversity by increasing speciation rates or the density of individuals and coexisting species; (2) factors that may limit diversity by increasing extinction rates, or by limiting individual density or niche space; and (3) factors that describe the role played by habitat area and complexity. Of these, theories relating diversity to environmental gradients in temperature, productivity, and habitat area have over time developed the most explicit mechanistic foundations.

Empirically, we also find that these three environmental predictors gather the most support across realms and species groups (table 3.6). When we combined the available information on primary and secondary environmental predictors analyzed at global or near-global scales (see table 3.6), some interesting patterns emerged. Across all 36 studied species groups discussed here, ambient temperature (or PET, which is closely related to temperature) was by far the most consistent primary predictor of species richness (72% of studies), whereas habitat area (6%) and productivity (14%) were less commonly identified as primary predictors. Habitat area emerged as the most important secondary predictor in 39% of cases, with temperature (14%) and productivity (14%) gathering less support.

TABLE 3.6. Environmental Predictors of Species Richness

Habitat	Species group	Primary predictor	Secondary predictor	Source
Coastal	Stony corals	Temperature	Habitat area	Tittensor et al. 2010
Coastal	Brittle stars	Temperature	Productivity	Woolley et al. 2016
Coastal	Bivalves	Temperature	Productivity	Valentine and Jablonski 2015
Coastal	Cone snails	Temperature	Habitat area	This volume, table 3.1
Coastal	Cephalopods	Temperature	Habitat area	Tittensor et al. 2010
Coastal	Benthic macrofauna	Temperature	Chlorophyll (−)	Macpherson 2002
Coastal	Seagrasses	Temperature	Habitat area	Tittensor et al. 2010
Coastal	Mangroves	Temperature	Habitat area	Tittensor et al. 2010
Coastal	Macroalgae	Temperature	Nutrients	Keith et al. 2014
Coastal	Fish	Temperature	Habitat area	Tittensor et al. 2010
Coastal	Sharks	Habitat area	Temperature	Tittensor et al. 2010
Coastal	Pinnipeds	Temperature (−)	Productivity	Tittensor et al. 2010
Pelagic	Foraminifera	Temperature	Habitat area	Rutherford et al. 1999; Tittensor et al. 2010
Pelagic	Copepods	Temperature	Salinity	Rombouts et al. 2009
Pelagic	Euphausiids	Temperature	Habitat area	Tittensor et al. 2010
Pelagic	Squids	Temperature	Habitat area	Tittensor et al. 2010
Pelagic	Macrofauna	Nitrate	Temperature	Macpherson 2002
Pelagic	Tuna and billfish	Temperature	Habitat area	Worm et al. 2005; Tittensor et al. 2010
Pelagic	Sharks	Temperature	Habitat area	Tittensor et al. 2010
Pelagic	Cetaceans	Productivity	Temperature	Tittensor et al. 2010
Pelagic	Seabirds	Habitat area	Temperature (+/−)	Davies et al. 2010
Pelagic	Bacteria	Temperature	—	Pommier et al. 2007; Fuhrman et al. 2008
Pelagic	Microbes	Temperature	—	Sunagawa et al. 2015
Deep sea	Brittle stars	Export productivity (+/−)	Oxygen stress	Woolley et al. 2016
Deep sea	Gastropods	Export productivity (+/−)	Depth	Tittensor et al. 2011
Deep sea	Bivalves	Export productivity (+/−)	Temperature	Tittensor et al. 2011
Deep sea	Foraminifera	Export productivity	—	Culver and Buzas 2000
Land	Vascular plants	Potential evapotranspiration	Wet days	Kreft and Jetz 2007
Land	Soil fungi	Moisture	Soil calcium	Tedersoo et al. 2014
Land	Soil microbes	pH (+/−)	—	Fierer and Jackson 2006
Land	Amphibians	Temperature	Productivity	Belmaker and Jetz 2011
Land	Birds	Temperature	Productivity	Belmaker and Jetz 2011
Land	Mammals	Temperature	Elevational range	Belmaker and Jetz 2011
Land	Vertebrates	Temperature	Habitat area	Jetz and Fine 2012
Freshwater	Vertebrates	Temperature	Habitat area	Tisseuil et al. 2013
Freshwater	Crayfish	Temperature	Habitat area	Tisseuil et al. 2013

Predictor	Main predictor	Secondary predictor
Temperature	26	5
Habitat area	2	14
Productivity	5	5
Other	3	8

Note: All effects are positive, except where indicated: (−) denotes a negative correlation with richness, (+/−) a unimodal one. Cumulative evidence for the three most common predictors is summarized at the bottom of the table. All publications included here were global in scope and data coverage, except for five studies that were at the scale of ocean basins (Culver and Buzas 2000; MacPherson 2002; Rombouts et al. 2009; Tittensor et al. 2011) or continents (Fierer and Jackson 2006).

Overall, ambient temperature was a primary or secondary predictor of global species richness in 31 of 36 cases (86%). Exceptions are deep-sea taxa, in which richness is related to export productivity, and soil microbes, and fungi, for which richness relates to soil chemistry and moisture (see table 3.6). It is important to note that in studies where colinearity of temperature with latitude, oxygen concentration, or seasonality was accounted for, the effects of temperature always remained significant. Thus, temperature and habitat together explained most of the patterns, whereas productivity-related variables were important for deep-sea species and marine mammals (see table 3.6).

Factors other than temperature, habitat, and productivity were rarely identified as primary predictors (5%) but sometimes as secondary correlates of species richness (22% of cases; see table 3.6). Here, species-specific habitat variables such as salinity, elevational range, or soil calcium were sometimes significant, showing that other factors can play important roles. Despite this, the overall consistency of our results raises the question of why there has been so much dissent and controversy regarding the drivers and predictors of species richness gradients in the past (Rohde 1992). It appears that global data coverage across multiple species groups, in combination with modern geospatial analysis methods, now allows for a much stronger test of environmental predictors than previous analyses of latitudinal gradients. Such comprehensive tests often produce surprisingly consistent results with common environmental drivers across diverse taxonomic groups (McCoy and Heck 1975; Roberts et al. 2002; Jetz et al. 2009; Tittensor et al. 2010; Belmaker and Jetz 2011; Tisseuil et al. 2013).

We conclude that there is good evidence for the effects of thermal energy (as a primary driver) and habitat area and productivity (secondary drivers) for explaining global richness patterns across different habitats and species groups (see table 3.6). This conclusion is supported by empirical analysis of present-day patterns and trends, by careful examination of the fossil record, and by ecological theory. In the next chapter, we will explore how these empirical insights can be combined into a mathematical theory and testable model of global biodiversity.

CHAPTER FOUR

Developing a Theory
of Global Biodiversity

In the previous chapters, we concluded that large-scale patterns of species richness appear to converge for well-studied species groups within each of the major biogeographic realms (terrestrial, coastal, pelagic, and deep sea). We further showed that there is some consistency to the environmental predictors of species richness when analyzed at a global scale, and that these predictors are linked to specific drivers and mechanisms. In particular, thermal energy emerged as the major predictor across taxa, with measures of habitat area and productivity also being significant across many species groups. Collectively, these results are most consistent with the *evolutionary speed hypothesis* (Stehli et al. 1969; Rohde 1992; Allen et al. 2002) and the *more-individuals hypothesis* (Hutchinson 1959; Wright 1983; Clarke and Gaston 2006).

In this chapter, we will develop a body of theory that allows us to capture and test the previously described key processes governing the global distribution of biodiversity. From this theory, we devise a spatial metacommunity model that enables us to reconstruct documented patterns of species richness from first principles, and to predict their major features. Specifically, in this model, the patterns must emerge from a combination of measurable ecological and evolutionary processes, rather than from a statistical fit to environmental correlates. As such, this is a mechanistic, process-based model *sensu* (Hilborn and Mangel 1997), although we do not endeavor to incorporate all possible mechanisms or processes right away. Instead, we chose to start with a simple, flexible, and tractable framework that can be built on and expanded in order to test competing hypotheses. This modeling approach may be described as an experimental toolbox for global biodiversity patterns. Our aim is not necessarily to achieve the highest predictive power, but to explore the possibility space of global biodiversity patterns and their drivers. As such, the model and its underlying theory are "instrumental" rather than "realistic" (Wennekes et al. 2012), and hopefully a useful foundation for others to improve and challenge. We are motivated by a recent call for an expansion of theory in ecology to "accelerate scientific progress, enhance the ability to address environmental challenges, and foster the development of synthesis and unification" (Marquet et al. 2014). We aim for an "efficient theory," as defined by Marquet et al. (2014)—that is, a theory that is grounded in first principles and mathematical expressions, makes few assumptions, and generates a large number of predictions per free parameter, enabling the testing of model predictions against multiple empirical datasets.

The building blocks for our theory of global biodiversity are based on the empirical evidence assembled and analyzed in the previous chapters. From our review of the existing literature on marine and terrestrial biodiversity patterns, we surmise that the primary forces that structure large-scale patterns of species richness are most likely related to some combination of ambient thermal energy, habitat area, or productivity. The effects of thermal energy can be captured through metabolic equations relating rates of community turnover and evolution (that is, evolutionary speed) to temperature. The effects of habitat area and productivity can both be captured through equations linking species richness to *community size* (that is, the number of individuals in a community). The equations describing these theoretical building blocks and resultant models are outlined in the rest of this chapter.

We begin with a simple, spatially explicit null-model built on Hubbell's well-known metacommunity model that implements his neutral theory of biodiversity and biogeography (Hubbell 2001). This model was chosen as a base because it incorporates both evolutionary processes of speciation and extinction and ecological processes of dispersal and disturbance, all of which influence species richness (Vellend 2010; Tittensor and Worm 2016). Like its predecessor, the theory of island biogeography (MacArthur and Wilson 1967), this neutral model assumes ecological equivalence among species. This assumption is of course an unrealistic abstraction of nature. Yet neutral theories have performed reasonably well in predicting some major macroecological properties such as patterns of relative abundance, range size, and diversity, although many open questions remain (Bell 2001; Rosindell et al. 2012; Marquet et al. 2014). This suggests at least the possibility that species-specific differences may not always be critical in driving major macroecological patterns. We evaluate this proposition with respect to observed global patterns of species richness across taxa. Subsequently, we relax that assumption by departing from neutrality and allowing species within our spatially explicit model to differ in their niche width and thermal tolerances, accounting for observed differences among species in nature (Sunday et al. 2011; Beaugrand 2014). In this way, we are combining elements of neutral theory, metabolic theory, and niche theory into a synthetic theory of global biodiversity.

4.1. BASIC NEUTRAL THEORY

Our theory, building on Hubbell (2001), incorporates the fundamental processes that drive species change in a metacommunity, specifically the birth and death of individuals, dispersal to and from neighboring communities, as well as the evolutionary processes of speciation and extinction. Hubbell's ideas were first implemented in a simple, nonspatial metacommunity model where a local community of n individuals corresponded (via immigration) with a larger, "external" species pool. Later versions have modeled that species pool as a spatially explicit metacommunity, in which many local communities are linked to each other via dispersal across a model landscape (Bell 2001; Hubbell 2001; Rosindell et al. 2008; Rosindell et al. 2011). This model structure allows ecologists to investigate spatial processes explicitly. On the downside, such a spatially explicit version is analytically tractable only for the very simplest of cases (see box 4.1). More complex scenarios, like the global richness patterns examined in this book, require numerical simulation models (Bell 2001; Hubbell 2001). Such models can be very computationally expensive, as every individual is tracked for every time step across a complex metacommunity, potentially for many thousands of time steps, until the model settles to a dynamic equilibrium. Even identifying when such models have achieved equilibrium can be challenging (Rosindell et al. 2008). Fortunately, a backward-simulation, or *coalescence* approach, partly alleviates these computational problems (Rosindell et al. 2008), as explained later.

BOX 4.1

ANALYTICAL SOLUTION FOR A SPATIALLY EXPLICIT NEUTRAL MODEL

Here, we briefly demonstrate that beyond a trivially simple system, a spatially explicit metacommunity model cannot be solved analytically and that a simulation approach must be used. Adopting an approach analogous to that of Hubbell (2001, 86), we can calculate the transition probabilities for species in a local community, assuming exactly one death and one birth per time step.

In the simple case of no metacommunity dispersal from neighboring cells ($m = 0$), we have a set of unlinked local communities, and can use the Markovian matrix approach of Hubbell (2001, 79). The species in each community reach an absorbing state of either monodominance or extinction, with the probabilities of each following exactly from Hubbell (2001), and with a vector of fixation times calculated by multiplying the fundamental matrix (as denoted therein) by an identity column vector. In the case of one birth and one death per time step, as per our model, the ith element in this fixation vector is

$$T(N_i) = (J-1)\left[(J-N_i)\sum_{k=1}^{N_i}(J-k)^{-1} + N_i\sum_{k=N_i+1}^{J-1}k^{-1}\right],$$

where J is the size of the local community, and N_i is the number of individuals in species i (Hubbell 2001).

(Box 4.1 continued)

When the communities are linked via dispersal and immigration, we can write out the transition probabilities for the ith species (of local community abundance N_i) in the ergodic community as:

$$\Pr\{N_i - 1 \mid N_i\} = \frac{N_i}{J}\left[m(1-P_i) + (1-m)\left(\frac{J-N_i}{J-1}\right)\right]$$

$$\Pr\{N_i \mid N_i\} = \frac{N_i}{J}\left[mP_i + (1-m)\left(\frac{N_i-1}{J-1}\right)\right]$$

$$+ \left(\frac{J-N_i}{J}\right) \times \left[m(1-P_i) + (1-m)\left(\frac{J-N_i-1}{(J-1)}\right)\right]$$

$$\Pr\{N_i + 1 \mid N_i\} = \left(\frac{J-N_i}{J}\right)\left[mP_i + (1-m)\left(\frac{N_i}{J-1}\right)\right],$$

where $0 <= m <= 1$ is the probability of immigration from the metacommunity, and P_i is the relative proportion of individuals in the metacommunity (here, the Moore neighborhood) that belong to species i (Hubbell 2001).

However, in our case, the metacommunity is not a separate, larger pool of individuals, but constitutes the set of surrounding local communities. These surrounding communities also have their own (overlapping) metacommunities. Thus, all local communities are simultaneously part of multiple metacommunities for other local communities, and this dependence brings about a recursive aspect to the system, whereby P_i becomes contingent.

For example, consider a very simple system of just two local communities, A and B. Assume that local community B acts as the metacommunity for local community A, and A acts as the metacommunity for B. For community A, the P_i terms in the preceding transition probabilities become P_B, the relative proportion of individuals in community B. And, similarly, for community B, the P_i terms become P_A. So the dynamics of B depend on A, which depend on B, and so forth. Thus, there is a two-way coupling of dispersal between communities, not just a one-way flow of individuals from the metacommunity to local communities. Deriving the elements in the transition matrix (not shown), as well as being extraordinarily complex for anything but a trivial spatial system, demonstrates this inherent recursiveness. An expression for the multinomial probabilities of individual states, and hence a Markovian process, cannot be written in fully explicit terms, and hence we cannot utilize the established machinery for calculating the unconditional probability of every possible configuration of relative species abundance in a sample of arbitrary size drawn at random from the metacommunity (Hubbell 2001). Instead, we rely on simulation approaches implemented either in forward or coalescence mode.

Two fundamental principles underlie the basic model implementation. The first principle states that all processes that are parameterized (disturbance, dispersal, and speciation) are modeled as per capita rates, such that all of these processes have a given probability per individual. Note that species extinction is not parameterized, but emerges from the model due to the random dynamics of births and deaths. All individuals regardless of species are assumed *ecological equivalents*— that is, none are more likely to be disturbed, disperse, or speciate than any other, and are selected at random to undergo these processes. We also assume *zero-sum ecological dynamics*, meaning that there are a fixed number of individuals that can be supported by a given environment, and while there is constant disturbance rate and associated death and turnover of such individuals, their total number does not change. This type of behavior applies only to ecologically similar species at one trophic level—for example, metacommunities of trees, corals, herbivorous mammals, or planktivorous fish—which are likely to be competing with one another for resources such as space (Hubbell 2001). However, it is possible to envision overlaying multiple models for different trophic groups atop one another, in an attempt to determine broader biodiversity patterns, and perhaps to examine interactions between these groups. We will return to this point later in this book.

The second fundamental principle is that environmental parameters thought to influence and shape diversity patterns, such as ambient temperature, or habitat area, are added in stepwise fashion to the base model. The expanded model is still purely neutral within each local community, with per capita rates being identical among all individuals, but now includes spatially heterogeneous environmental factors across the metacommunity.

Our initial model begins from a local community that is composed of a number of sites (fig. 4.1), and has no connection to a metacommunity. Each of these sites is occupied by a single individual of a designated species. Local community size J is constant through time. The local community is initially seeded at random from a pool of S species, where each species has an equal chance of occupying each site, in accordance with the principle of neutrality. It was previously shown that the equilibrium community of this type of model was not sensitive to the initial number of species (Hubbell 2001). At each time step, one randomly selected site is disturbed, and the individual occupying that site dies and is replaced at random, where the probability of a particular species filling that site is proportional to its local abundance in that same community. Hence, species that attain a higher abundance in the community are more likely to recolonize disturbed sites in the next time step. Because of this positive feedback, the community converges to a monoculture ($S = 1$) over time. This result is well established for neutral models without immigration (Hubbell 2001).

Next, we add immigration to the model by linking the local community to a global, spatially explicit, metacommunity M, allowing for dispersal between local

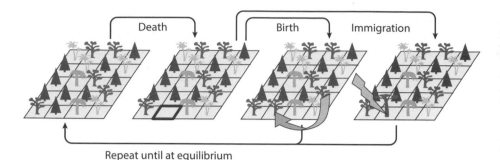

Repeat until at equilibrium

FIGURE 4.1. Basic mechanics of a neutral model community. A local community is occupied by *J* individuals belonging to *S* different species. A randomly selected individual dies at each time step (square symbol) and is replaced at random with an individual born in the same community (arrow symbol) or with an individual immigrating from a neighboring community (lightning symbol). The probability of immigration is given by the dispersal parameter *m*. The full model also allows for dispersing individuals to speciate as they colonize a neighboring community (presumably by point mutation and reproductive isolation). The probability is given by the speciation rate *v*. In a neutral community, the per capita rates of birth, death, dispersal, and speciation are assumed to be equal among all individuals, and subject only to random variation or drift. Redrawn after Rosindell et al. (2011).

communities with rate *m*. The global metacommunity (here assumed to represent a hypothetical planet-spanning global ocean) is composed of $i \times j$ local communities, where *i* is the number of local communities longitudinally, and *j* is the number latitudinally. The communities are assumed to be distributed on a regular grid over a cylinder—that is, cells are connected over the longitudinal borders at the 180th meridian, but not the poles. Fig. 4.2 illustrates that basic spatial structure. Each local community is connected via dispersal to its eight nearest local community neighbors, the *Moore neighborhood* of cellular automata theory (Hubbell 2001). Polar cells on the top or bottom row of the grid instead have a neighborhood size of five local communities, since there is an upper or a lower border through which species cannot disperse. As per the zero-sum assumption, the total number of individuals in the global metacommunity (J_M) is constant. It is given by the product of the local community size (*J*) multiplied by the number of local communities ($i \times j$).

At each time step, a single site within the entire metacommunity is disturbed and the individual that occupied that site dies and is replaced, either from the local community or from a neighboring community. This approach settles to the same dynamic equilibrium as disturbing one site within every local community per time step (Hubbell 2001; Tittensor and Worm 2016), though it takes many more time steps to reach a stable equilibrium. However, the advantage

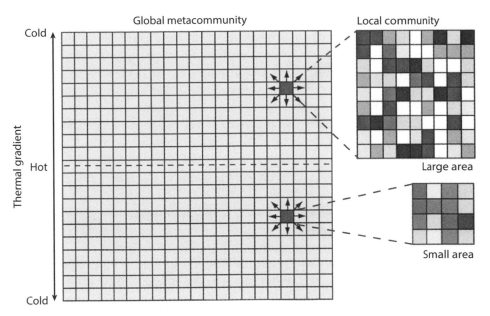

FIGURE 4.2. Global metacommunity model. Our spatially explicit global metacommunity consists of 21 × 21 local communities, each occupied by J individuals (with local community size $J = 16$ and metacommunity size $J_M = 7056$ as the base case). Shades in local communities symbolize different species. Communities are connected through dispersal (at rate m) to neighboring cells (arrows). Although the grid is depicted as flat, the left-hand and right-hand borders are connected to form a cylindrical shape. The dashed line in the center represents the equator, and the top and bottom of the main grid represent the poles. Disturbance and replacement of individuals on the grid is a per capita function that is constant for all local communities (in the neutral base model) or can become a function of temperature (in the combined neutral-metabolic model), with increased turnover rates or speciation rates at higher temperature. The effect of increasing habitat area or productivity can be simulated by scaling the number of individuals J in any specific local community up or down. Redrawn after Tittensor and Worm (2016).

of selecting a single disturbance across all communities is that this particular method is more easily reconfigured for a nonuniform probability of disturbance between communities (Tittensor and Worm 2016). This approach is implemented later, in which local communities have different probabilities of disturbance based on environmental conditions.

When a site gets disturbed, the individual occupying it is replaced at random with an individual dispersing from a neighboring community (the Moore neighborhood) with probability m, or from another site within the local community with probability $1 - m$. The probability of a specific *species* occupying a disturbed site is proportional to its relative abundance in the Moore neighborhood in the first instance, and in the local community in the second. At each time step, an individual can move only to immediately neighboring communities. Ultimately, however, it can spread through the global community via successive Moore neighborhoods (Bell 2001), with faster dispersal through the metacommunity at higher values of m.

In this simple metacommunity model, species richness declines over time to monodominance, due to the random processes outlined earlier, albeit more slowly than for a single community. We find that even a low dispersal rate ($m = 0.01$) introduces a spatial *rescue effect* that slows the spread of local monocultures (and decelerates global extinction). Higher dispersal rates initially lead to higher average local species richness in each community, but lower global richness across the metacommunity, as found by Hubbell (2001, 218). Ultimately, however, without the input of any new species to the system, it will eventually settle to a monodominant system. Because species are ecologically identical, the species that comes to dominate cannot be predicted a priori.

Next, we introduce speciation into the model, which has a chance of occurring when a new site is colonized by a dispersing propagule. Thus, every speciation event corresponds to a dispersal event: dispersal occurs and then speciation is determined independently. Speciation is assumed to occur as a point mutation with a per capita probability of v. Although other speciation mechanisms have been proposed (Rosindell et al. 2010), we use this simplest possible form in our first attempt at a simple process-based model. Experimenting with different forms of speciation remains one of many aspects that could be tested within the model framework.

Neutral theory characterizes the (sampling) distribution of species abundances in a metacommunity of size J_M and speciation rate v by a composite parameter known as the fundamental biodiversity number,

$$\theta = J_M \frac{v}{1-v}$$

(Hubbell 2001; Rosindell et al. 2011). A number of well-known ecological properties emerge from a spatially explicit metacommunity model of this form, including the species-area relationship (Hubbell 2001, 152–201), patterns of species' ranges (see Hubbell 2001, 216–217), and rank-abundance curves (Hubbell 2001, 219), among others. Note that these patterns emerge entirely due to random processes parameterized by the speciation rate and dispersal parameter, not because of any structure in the environment, or any assumed variation in species' traits.

Our model therefore entails the following basic steps (figs. 4.1 and 4.2):

1. At every time step, a randomly selected community in the global metacommunity has a random site disturbed and the individual residing therein dies.

2. The empty space is then colonized by an individual from the metacommunity with probability m (dispersal parameter) or from the local community with probability $(1 - m)$, with the individual chosen at random.
3. The newly colonizing individual evolves into a new species with probability v (speciation parameter).
4. The preceding steps are repeated until a dynamic equilibrium is reached (see section 4.2, later).

In summary, this basic model implementation of spatially explicit neutral theory makes the following assumptions:

1. Dispersal, mortality, and speciation rates are constant and individuals are ecologically equivalent (identical per capita rates regardless of species identity).
2. The number of vacant sites opening up in a community is proportional to the disturbance rate, which is equivalent for each local community, as one random site in the metacommunity is disturbed per time step.
3. The probability of a species filling a single vacancy is equal to its relative abundance in the local community (if local dispersal) or Moore neighborhood (if metacommunity dispersal).
4. A finite amount of space is assumed in each local community (J) with a constant number of individuals in the metacommunity, J_M, competing for the space across all local communities.

This model differs from the spatial model described in Hubbell (2001) and elsewhere only in the implementation, in which a single community is disturbed each time step, rather than all communities being disturbed in each time step. This produces identical results to these other model implementations, but will become important later, when additional components are added to the model, and assumptions of neutrality are relaxed.

4.2. IMPLEMENTATION IN FORWARD AND COALESCENCE MODE

Computationally, the model is implemented using two different algorithms or *modes*, each of which has advantages and disadvantages, but which can be used as independent checks on one another. The first is *forward mode*, which is the simulation approach used in Hubbell (2001). It is used primarily to examine and visualize transient and nonequilibrium community dynamics, and proceeds to simulate community changes under a given set of parameters until a finite number of time steps or community turnovers have been completed. The number of time steps necessary to reach a dynamic equilibrium is dependent on the speciation rate.

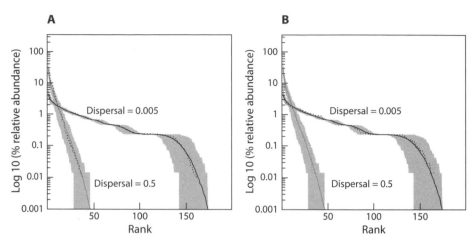

FIGURE 4.3. Comparison of the global metacommunity model with original results by Hubbell (2001). Rank-abundance plots are depicted. (A) Model results for a metacommunity of 7056 individuals with fundamental biodiversity number $\theta = 10$ simulated in coalescence mode (see text). Solid lines indicate mean values, and the gray-shaded area indicates 95% range of 100 replicate runs. Dashed black lines indicate values from fig. 7.7 of Hubbell (2001, 219) at the same parameter settings. (B) Same, but solid lines represent results from forward-simulation mode, run to 25,000 community turnovers, for approximately 1.7 billion time steps. Both forward and coalescence modes produce results identical to Hubbel (2001).

Hubbell (2001) states that approximately 10,000 community turnovers (equivalent to each individual site being disturbed 10,000 times, on average) should suffice, but we found that at low speciation rates, the model needed to be run for longer to settle to a dynamical steady-state. We therefore assessed settlement to a dynamic equilibrium based on the settling of transient properties (for example, relative species abundances), and on comparison to the equivalent coalescence model explained later (see also fig. 4.3). The metacommunity is initially seeded with a single species; we note that the transient dynamics are dependent on the initial number of species (low initial richness gives an increase in richness, high initial richness a decrease), but ultimately all initial conditions settle to the same equilibrium (as noted by Hubbell 2001). We choose to start with a single species, as we believe it more accurately reflects the process of evolution in novel environments, but have checked our results starting with high species richness, and will indicate where any differences in transient dynamics occur.

As a second approach to model implementation, *coalescence mode* (Rosindell et al. 2008) allows us to run large numbers of simulations in a computationally tractable manner by beginning at the end-state and working backward without having to compute redundant events that have no bearing on the final distribution of species. Coalescence algorithms employ techniques originally developed to track neutral evolutionary markers in population genetics (Felsenstein 2004) and offer a number of important advantages over traditional simulation approaches. In this instance, coalescence begins with a present-day global metacommunity where each individual is of unknown species identity and lineage, and then works backward in time, applying the rules of the model in reverse and probabilistically determining speciation and dispersal events that can uniquely identify individuals as being derived from the same lineage (Rosindell et al. 2008). This mode efficiently uncovers the ancestry and phylogeny of contemporary individuals in the sample, but without modification can be applied only to strictly neutral models. Crucially, it provides a way to simulate our individual-based theory for large and complex spatial structures; no known alternative algorithm would be tractable in these situations (Rosindell et al. 2008; Rosindell et al. 2011). This gain in speed and efficiency occurs through tracking only individuals whose lineage has not yet been determined, and not those whose identities have been established, whereas the forward model continues to follow every individual during all time steps. In our tests, coalescence mode generally yields the same results as the more traditional forward approach (see fig. 4.3) but is up to 1,000 times faster, allowing us to perform replicate runs that can be used to average over the stochastic variation inherent to each simulation. Coalescence also provides an objective end point, unlike the forward mode, for which it can sometimes be challenging to determine that the model has fully "settled" (Rosindell et al. 2008). The coalescence endpoint is reached when the identity of each species has been established, giving a complete picture of the present state based on the ecological and evolutionary events that have occurred. Owing to its computational efficiency, the coalescence approach allows us to run many replicate simulations. In this chapter, unless otherwise noted, we typically run 100 simulations for each parameter combination and present the mean values and 2.5% and 97.5% percentiles.

One beneficial aspect of creating two almost entirely separate algorithms was that each could be used as a check on the other. This redundancy promotes greater confidence in our results, and it forced us to think clearly about how the implemented ecological and evolutionary processes operated, given that each had to be interpretable and specifiable both in forward and coalescence mode.

To enable direct comparison with Hubbell's spatial metacommunity model, we initially set the size of each local community as $J = 16$, and the number of communities in longitudinal and latitudinal directions, i and j, to be 21, for a total metacommunity size of $J_M = 7056$ individuals (see fig. 4.2). While this relatively tractable metacommunity is used in most simulations, we explored larger communities in our sensitivity analyses. We further explored a range of values for the speciation rate v, ranging from 0.0001 to 0.1, and for the metacommunity dispersal rate m, from 0.001 to 1. Furthermore, for direct comparison to Hubbell (2001), we ran an analysis in which the dispersal parameter is set to either 0.005 or 0.5, and the speciation rate to 0.000709, giving a fundamental biodiversity number (θ) of 10. Results in both forward and coalescence mode were indistinguishable from Hubbell's original model (see fig. 4.3). However, note that unlike Hubbell, we did not embed the 21×21 grid of communities in a large plane, though longitudinal edge effects will be avoided in our model, as the metacommunity grid is linked along this axis.

The basic model described here does not produce any gradients in biodiversity due to the lack of geographic processes that may structure diversity between regions. For this, we need to introduce a factor or combination of factors that vary geographically across the global metacommunity. Based on the empirical findings of chapter 3, we first focus on the gradient in ambient temperature, thereby integrating metabolic and neutral theory into a unified theoretical framework.

4.3. INCLUDING METABOLIC THEORY

As discussed in the previous chapter, increases in ambient temperature are known to increase individual metabolic rates, and consequently the rate of species interactions and community turnover (Brown et al. 2004). This forms a first link in our theory between a hypothesized driver (thermal energy), a documented mechanism (metabolic rates), an empirical predictor (temperature), and observed patterns of species richness. We simulate this driver within our model by establishing a temperature gradient from the equator to the poles of our global metacommunity, and making the rate of disturbance and recolonization temperature dependent based on metabolic rates (Tittensor and Worm 2016). Although this remains, *sensu stricto*, a neutral model in that all individuals are ecologically equivalent, with identical per capita rates within each local community, the rates for each local community are dependent on the latitudinal position of those communities. We further assume that individual metabolism and turnover rates speed up at higher temperatures, and that this effect is the same for all species. Clearly, this model is more applicable to ectotherms than endotherms. Purely as a result of faster turnover, we also expect to observe more speciation events per unit of time at lower latitudes. Thus, we induce an environmental gradient without changing the basic neutrality assumptions of per capita equivalence (see fig. 4.2).

The effects of temperature on community turnover and speciation rate are parameterized according to the metabolic theory of ecology (Brown et al. 2004) described in the previous chapter. According to the theory, biochemical reaction rates, metabolic rates, and other rates of biological activity increase exponentially as a function of internal organism temperature and body size (see chapter 3). In the context of our neutral model, we can ignore the effect of body mass, as all organisms are assumed ecological equivalents. The temperature dependence is described by the Boltzmann factor $e^{-E/kT}$, where E is the empirically derived activation energy (~0.65 eV for ectotherms, as stated by Brown et al. 2004), k is Boltzmann's constant (8.617×10^{-5} eV K^{-1}), and T is absolute internal temperature in degrees Kelvin (K). This relationship predicts about a doubling of metabolic rate with every 10 K increase in temperature. We assumed a 30 K gradient in average surface temperature between the equator and each pole, approximating the observed gradients in sea surface temperature in today's oceans (fig. 4.4A). This realistic temperature gradient results in an approximately eightfold difference between average metabolic rates from the equator to the poles for ectotherms. This process was simulated in the model by making the probability of disturbance at each site in a local community proportional to the rate difference (Tittensor and Worm 2016). Therefore, a site located in an equatorial community had around an eightfold higher chance of being disturbed in any individual time step than one at the poles, with intermediate latitude cells following a gradient between the two. We always normalized the total probabilities to sum to one, to enable the simple random selection of a cell for disturbance.

When we implemented the specified 30 K latitudinal temperature gradient along with its predicted effect on community turnover in the forward model, we observed an unstable latitudinal gradient in mean local community species richness (fig. 4.5). Faster turnover and shorter generation times in the tropics resulted in a more rapid approach toward the dynamic equilibrium richness state

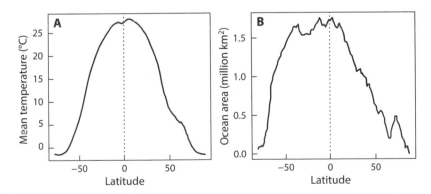

FIGURE 4.4. Observed global gradients in sea surface temperature and total ocean area by latitude. (A) Surface temperature and (B) ocean area both peak around the equator and decrease sharply toward the poles. The regional peak of ocean area in the Arctic partly reflects areas permanently covered by ice. Redrawn after data in Tittensor et al. (2010); Allen and Gillooly (2006).

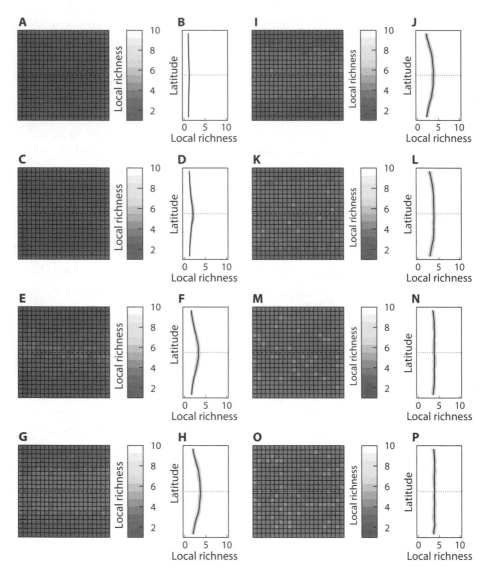

FIGURE 4.5. A transient biodiversity gradient resulting from increased community turnover at higher ambient temperature. Left-hand column depicts a global metacommunity grid of 21×21 communities with shading indicating their species richness S. Right-hand column depicts the resulting latitudinal gradient, with the solid line indicating the mean, and shading representing the 95% range of 100 model runs. Shown is the appearance of a transient latitudinal gradient in species richness from a forward-mode neutral-metabolic model after (A, B) 1 community turnover (7056 time steps); (C, D) 10 community turnovers; (E, F) 50 community turnovers; (G, H) 100 community turnovers; (I, J) 125 community turnovers; (K, L) 250 community turnovers; and (M, N) its disappearance within 1000 community turnovers. Coalescence model results (O, P) are shown for comparison. Dispersal parameter $m = 0.1$; speciation rate $v = 0.01$; global metacommunity size $J_m = 7056$ individuals. After Tittensor and Worm (2016).

there, elevating tropical relative to polar richness. Eventually, however, the polar regions did catch up and reached the same asymptotic richness and relative species abundance distribution as equatorial areas (fig. 4.5M,N). The coalescence-mode simulation (fig. 4.5O,P) confirmed that no gradient existed at equilibrium. This partly resulted from the lack of any stabilizing force to maintain the asymptotic gradient of equilibrium richness—it is simply a matter of time—and partly from the initial richness being lower than equilibrium richness. This latter point can be demonstrated by starting a forward-mode model with local richness values set to be higher than the mean equilibrium value, in which case the more rapid approach of equatorial regions toward the equilibrium value produces a transient inverse latitudinal gradient (results not shown). We note that it actually remains an open question as to whether Earth has reached an equilibrium in terms of global biodiversity, a fact that is deemed unlikely given the continuing increase of fossil biodiversity up to the present, as well as the large-scale disturbance events that "reset" biodiversity at regional and sometimes global scales (Raup and Sepkoski 1982; Benton and Twitchett 2003; Alroy et al. 2008).

Metabolic theory also invokes an effect of temperature on speciation rates independent of the previously described effect on turnover times. As discussed in chapter 3, there is empirical evidence that mutation rates increase as a function of temperature (Muller 1928; Lindgren 1972; Matsuba et al. 2013). This may be mechanistically explained by the increased production of free oxygen radicals at higher metabolic rates, which increases nucleotide substitution rates, and may ultimately drive up speciation rates (Allen et al. 2006). However, we note that the mechanisms that link nucleotide substitution to speciation are not entirely clear. Despite this gap in mechanistic understanding, increased speciation rates during periods of climate change and increased temperature have also been evident in the fossil record (Stehli et al. 1967; Sepkoski 1998; Mayhew et al. 2012). We simulated such thermal effects on speciation rates in our model by setting the speciation rate, v, to be a normalized function of temperature:

$$v_j = \frac{v_{base}\, e^{-E/kT_j}}{\min(e^{-\frac{E}{kT_j}})},$$

where v_{base} is the "basic" speciation rate at the lowest temperature, and the metabolic effect of temperature, e^{-E/kT_j}, is normalized by dividing by its minimum value. Thus, at the poles the speciation probability equals v_{base}, and it increases exponentially toward the equator as predicted by metabolic theory (Brown et al. 2004). This effect can be simulated independently of or in combination with the effect of temperature on turnover rates.

We implemented this effect to explore under which conditions a permanent richness gradient emerges (fig. 4.6). As previously discussed, no effect was seen in the base model (fig. 4.6A,B), or when modeling the effects of temperature on

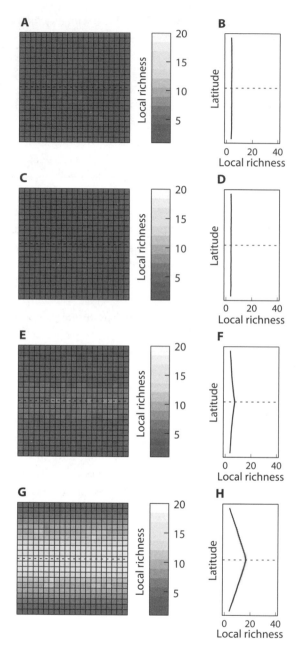

FIGURE 4.6. Emergence of permanent latitudinal biodiversity gradient. Shown are results from 100 coalescence model runs each: (A, B) in a neutral base model; (C, D) including a thermal effect on community turnover; (E, F) including thermal effect on speciation rates; (G, H) including a gradient of increasing habitat area from the tropics to the poles. Symbols and base parameter settings as in fig. 4.5. After Tittensor and Worm (2016).

community turnover (fig. 4.6C,D). However, when a metabolic effect on speciation was applied to our base model, a permanent latitudinal gradient emerged (fig.
4.6E,F). The mean richness of a local community at the equator ($S = 7.3$ species)
was around 2.1 times the average at the poles ($S = 3.6$), which showed near-identical
richness as the base model ($S = 3.5$). The resulting latitudinal gradient was near-linear, with some variability caused by the stochastic nature of our model (fig. 4.6E,F).

4.4. INCLUDING HABITAT AREA AND PRODUCTIVITY

A further extension of our theory and associated model implementation accounts
for the effects of habitat area and productivity. Species-area theory, both in its original form (Preston 1962) and as formalized in the theory of island biogeography
(MacArthur and Wilson 1967), rests on the fundamental assumption that islands
with larger habitat area can support larger numbers of individuals, and larger average population sizes would reduce the rate of local extinction (see chapter 3).
Species-area theory was later extended to include productivity, which also affects
the number of individuals that a given environment can support (Wright 1983).
The effects that these factors might have on species richness have been formalized
as the species-energy hypothesis *sensu* (Wright 1983) or the more-individuals
hypothesis *sensu* (Evans et al. 2005). Thus, our theory also ought to account for
the fact that some local communities support higher numbers of individuals than
others. We implemented this in our model by varying local community size, while
keeping the per capita rates of ecological processes, and the temperature of that
environment, constant.

As a first approximation to the real-world ocean, we assumed a fivefold area
gradient from the tropics to the poles, which roughly corresponds to the observed
gradient when averaging habitat area by latitude across oceans (see fig. 4.4B).
Running the basic model with this fivefold gradient resulted in larger community
sizes at the equator and produced a stronger latitudinal gradient than that for either
effect of temperature (see fig. 4.6G,H), with mean values at the equator almost 4.4
times higher than at the poles. Higher species richness evolved in lower latitude
cells, in part as a simple effect of increased community size and sampling across
more individuals, and in part owing to a higher absolute probability of speciation
occurring within these larger communities because of proportionally more frequent disturbance and dispersal events (but with constant rates per capita).

When modeling the combined effects of observed area and temperature gradients on species richness (fig. 4.7), their effects were multiplicative, producing a
stronger gradient with a sharper peak (fig. 4.7E,F), and a mean equatorial richness
value ($S = 35.4$), which was 9.2 times higher than the polar value ($S = 3.9$). Such
synergistic effects were not seen for turnover and area (fig. 4.7C,D), or when

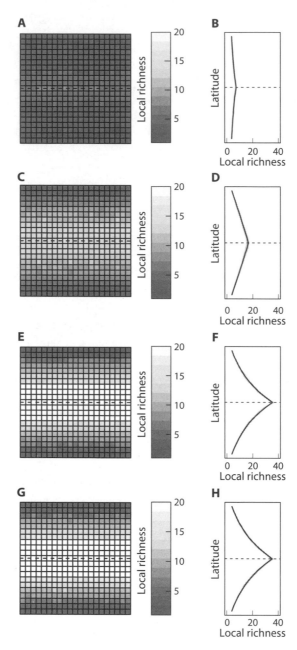

FIGURE 4.7. Combined effects of area and temperature gradients on latitudinal richness: (A, B) including effect of temperature on both community turnover rate and speciation rate; (C, D) including an area effect, and a temperature effect on community turnover rate; (E, F) including an area effect, and a temperature effect on speciation rate; (G, H) including an area effect, and a temperature effect on both community turnover rate and speciation rate. Symbols and base parameter settings as in fig. 4.5. Thermal effects are parameterized as per the metabolic theory of ecology. Area and temperature gradients approximate those found in the global oceans. After Tittensor and Worm (2016).

combining turnover and speciation and area effects (fig. 4.7G,H). We conclude that gradients in speciation rate and community size, but not turnover, produce persistent gradients in species richness. The effects of area and speciation can act synergistically, with model results roughly equally sensitive to variation in both parameters. Note that we assume here that increases in area-specific productivity may produce similar results as increases in habitat area (Wright 1983), as both are increasing local community size; we do not further distinguish between them here, but contrast them when fitting to data in chapter 5, accounting for the fact that they can show different scaling relationships with species richness in nature (Hurlbert and Jetz 2010).

We explored the sensitivity of metacommunity properties to variation in speciation rate v and dispersal parameter m for model runs with thermal gradient effects on both speciation and turnover (fig. 4.8). Mean local community species richness increased both with increasing metacommunity dispersal and with increasing speciation rates (fig. 4.8A). Clearly, higher speciation rates led to the evolution of more species, whereas higher dispersal rates allowed those species to spread over more local communities and avoid the higher extinction probability that plagues small isolated populations. Mean global metacommunity richness showed an even more pronounced increase with increasing speciation rate (fig. 4.8B), reflecting a more positive net balance between the probabilities of speciation and extinction. The effects of changes in the dispersal parameter m on global richness, however, were minimal, acting only to slightly increase the total richness at low dispersal values, probably owing to the greater "regionalization" of communities under low dispersal regimes, resulting in higher beta-diversity. When assessing the effect of both parameters on the strength of the latitudinal biodiversity gradient (fig. 4.8C), we saw an interaction between the effects of speciation rate and dispersal. While increasing speciation rate increased the strength of the latitudinal richness gradient across most tested dispersal parameter values, this effect was more pronounced at low dispersal rates than at high dispersal rates (see fig. 4.6). Dispersal had a slightly positive effect on gradient strength at low speciation rates, but a negative effect at the highest speciation rates. Note here that not all of these parameter combinations might occur in nature—for example, high dispersal rates might prevent speciation by undermining reproductive isolation, and thus high dispersal rates would not necessarily co-occur with high speciation rates. As a final sensitivity check, we explored the effects of changes in community size, and found that increasing community size to $J = 64$ ($J_M = 28{,}244$) individuals increased local and global richness at equilibrium but did not affect any of the latitudinal patterns we documented for smaller community sizes. While such parameter space exploration proved insightful, we acknowledge that the relationship to actual rates of dispersal and speciation, as well as actual community sizes found in nature, is not fully resolved and should be explored in future work.

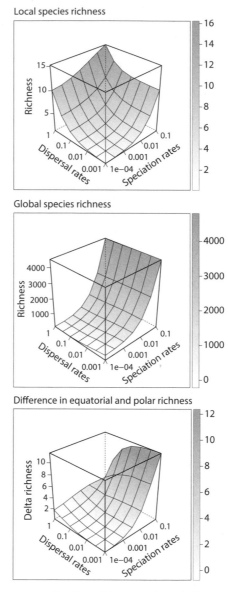

FIGURE 4.8. Parameter space exploration of the neutral-metabolic model. Community properties shown as a function of variation in dispersal and speciation rate parameters. (A) Effects on mean local community species richness. (B) Effects on global metacommunity species richness. (C) Effects on the strength of the latitudinal gradient, represented as the difference between mean equatorial and mean polar richness. Shown are mean values of 100 model runs. After Tittensor and Worm (2016).

4.5. INCLUDING TEMPERATURE NICHES

A final step in the development of this model framework introduces thermal tolerances and niches, and hence represents a departure from the strictly neutral model in that we can now assume ecological differences among species. In a sense, we are testing the sensitivity of previously described model results to the (very unrealistic) assumption of ecological equivalence by relaxing this assumption and evaluating the effect of thermal niches on observed patterns. For some marine species, it has been shown that we can reconstruct latitudinal ranges (Sunday et al. 2012) and some global diversity patterns (Boyce et al. 2008; Beaugrand et al. 2013) from individual species' thermal tolerances. Temperature niches have also been used extensively in habitat models and were invoked in a general model of latitudinal diversity, primarily applied to pelagic organisms (Beaugrand et al. 2013; Beaugrand 2014). Hence, we explored the effects of temperature niches on latitudinal patterns in our model.

We assumed that all species have the same niche width (w), but different midpoints in their thermal niche. The midpoint is given by the ambient temperature where the species evolved. For example, assuming a ±5 K niche width, a species that evolved in a 25°C environment is assumed to tolerate from 20 to 30°C ambient temperature. A species that evolved at 5°C may tolerate 0 to 10°C. Hence, species are still assumed equivalent in that they all have identical thermal tolerance ranges, but they also feature individual temperature niches in that they are adapted to different optimal temperatures. Within each local community, species remain equivalent, but within the metacommunity, they differ in their ability to disperse to regions of different temperature. Species are generally assumed to be unable to disperse outside their thermal niche. For phytoplankton, these assumptions have detailed empirical support. As shown from an analysis of all available data on temperature tolerance of phytoplankton strains, as well as from a mechanistic model, the temperature for optimum growth is closely correlated with the mean ambient temperature where the species is found (Thomas et al. 2012). Similar results have been described for reef fish and invertebrates (Stuart-Smith et al. 2015). Furthermore, the width of the temperature niche for phytoplankton species was independent of the mean ambient temperature of the environment where the species occurred (Thomas et al. 2012). These empirical findings support our simplified assumption of constant niche width. Finally, the maximum growth rate of individual phytoplankton strains increased with mean ambient temperature, indicating that adaptation to different temperature regimes does not override the general metabolic relationship between temperature and growth (Thomas et al. 2012). This means that high-latitude species have similar temperature niches as low-latitude ones, but with lower optimum temperature and lower realized growth rates, on average. This can be represented in our model framework by combining thermal effects on speciation and turnover with thermal niches.

Niches are implemented in the model in a very simple manner: should an individual attempt dispersal outside its thermal niche, it is assumed to have failed and the individual perished. In this case, a new dispersal event is drawn at the same location, retaining dispersal (m) and speciation (v) values (that is, whether a local or metacommunity dispersal, and whether a new species is formed) in order to maintain rates equivalent to the input parameter values. This approach is implemented equivalently in both forward-simulation and coalescence mode. In both approaches, when a new species evolves, the midpoint of its thermal niche is set to be equivalent to the temperature where it first appears. Thus, new species can evolve new thermal tolerances that deviate from their parent species, enabling a gradual successive colonization of species from the tropics to the poles—or vice versa.

The effects of niches on latitudinal ranges are displayed in figs. 4.9 and 4.10. Latitudinal range here has a theoretical maximum value of 21 latitudinal bands, corresponding to a pole-to-pole distribution in our standard 21 × 21 grid metacommunity (see fig. 4.2). In the simulations, we could see how niches constrained the maximum latitudinal ranges of species, and how progressively larger niches allowed species to realize greater ranges. A nonrestrictive niche of ±15 K spanned the entire temperature gradient of 30°C and led to an identical outcome as the base model without niches (fig. 4.9). It is very interesting, however, that the proportion of wide-raging species (here defined as those spanning more than 3 latitudinal bands) remained low irrespective of whether niches are present or not. The most restrictive niche ($w = \pm3$ K) had 99.9% of species at latitudinal ranges of 3 bands or less, the exceptions being species that spanned the equator. This dropped to 95% at the largest niche width and for the base case of no niches at all. In other words, at our standard parameter settings, only 1 out of every 20 species became wide ranging, whether niches were implemented or not; most species were wiped out before they could spread or survived only in smaller clusters of nearby communities. The maximum mean latitudinal range was 10 bands, which is much smaller than the theoretical maximum of 21 latitudinal bands (fig. 4.9).

The only case where we saw a marked effect of niches on the distribution of species occurred at the highest dispersal rates ($m = 1$; see fig. 4.10), when every dispersal event leads to colonization from a neighboring community. In this situation, which may approximate a mobile pelagic community, a larger fraction of species occupied expanded ranges, with the proportion increasing from 3 to 30% between a ±3 K and a ±15 K (nonrestrictive) thermal niche. Under this scenario, we also found that at least one species extended from pole to pole, reaching the theoretical latitudinal range maximum of 21 cells. Hence, it appears that, at least in our model, the constraining effects of niches on species distributions were

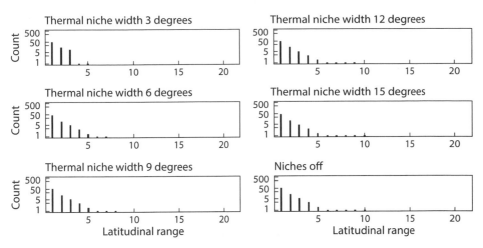

FIGURE 4.9. Neutral-metabolic-niche model under moderate dispersal. Shown are species latitudinal range distributions at dispersal rate $m = 0.1$, speciation rate $v = 0.01$, and under different
assumed thermal niche width ranging from ±3 to ±15 K. Note logarithmic scale on the y-axis.
Few species (up to 5%) colonize large ranges, irrespective of their thermal niche width. Note
that a thermal niche width of ±15 K is nonrestrictive given a total 30 K temperature gradient, and
produced identical results to a model run with niche parameter turned off.

realized only when there was no dispersal limitation. These results were upheld
at larger community sizes, and whether we allowed niches to evolve over time or
forced them to remain fixed (results not shown).

When we further explored the effects of niches on diversity patterns across
parameter space, we found that niches can have a positive effect (~50% increase)
on local diversity (fig. 4.11), due to greater *regionalization*. However, this was
seen only at the highest dispersal rates, and not at lower dispersal rates, where
high regionalization exists regardless of niche settings. Little effect was seen on
global metacommunity richness, but there was an effect on the strength of the latitudinal gradient at high dispersal rates (see fig. 4.11). Niches more than doubled
the steepness of the latitudinal gradient relative to the non-niche case at high
dispersal ($m = 1$), but not at low or moderate dispersal rates. These effects on latitudinal gradient strength were seen particularly at smaller niche width ($w = ±3$ K),
but less so at larger niche widths ($w = ±9$ K, $w = ±12$ K) (fig. 4.12).

These results make sense when considering latitudinal range size distributions
at different niche widths w and dispersal rates m (see figs. 4.9 and 4.10). Wide-
ranging species can "flatten" the latitudinal gradient, but only when they contribute a substantial proportion of overall species richness, which occurs only at high
dispersal rates (see fig. 4.10). Under these circumstances, niches are effective
in limiting the latitudinal spread of species, and hence in maintaining a steeper

116CHAPTER 4

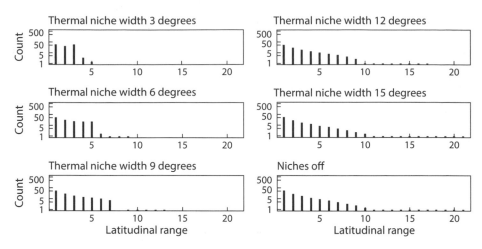

FIGURE 4.10. Neutral-metabolic-niche model under high dispersal. Shown are species latitudinal range distributions at maximum dispersal rate $m = 1$, speciation rate $v = 0.01$, and under different assumed thermal niche width ranging from ±3 to ±15 K. Under assumed maximum dispersal, a larger fraction of species (up to 30%) colonizes large ranges, with higher fractions at larger thermal niche width.

richness gradient. It is striking however, how little effect niches had at medium to low dispersal rates (figs. 4.9 to 4.12). It is interesting in this context that most of the empirical (Boyce et al. 2008; Thomas et al. 2012) and theoretical work (Beaugrand et al. 2013; Beaugrand 2014) on temperature niches affecting global diversity patterns has been conducted for pelagic communities that are typically very mobile, and have little dispersal limitation. Empirically, we see clear evidence that pelagic organisms tend to have larger ranges than their coastal counterparts (see fig. 4.13, for example). This could potentially affect their patterns of species richness, and partly explain the broad peaks of tropical to subtropical richness we observe in many pelagic taxa (see chapter 2).

When we check these results with our forward-simulation model, we find very similar patterns, but in addition we observe some transient effects of niches under most parameter combinations. Specifically, tropical richness was elevated when (restrictive) niches were present for 100 to 250 complete metacommunity turnovers, and a notable gradient appears after about 500 turnovers. Yet, over time, these effects disappeared, and patterns at equilibrium were near-identical to the case with no niches (results not shown).

At this point, we should clarify that of course niches can have large and important effects on individual species, their distribution, local abundance, and extinction risk (fig. 4.14). They also have clear effects on community composition and species identity, and may keep many species bound at the latitude at which they

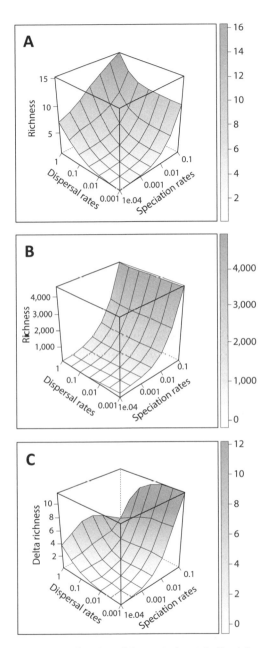

FIGURE 4.11. Parameter space exploration of the neutral-metabolic-niche model with assumed thermal niche width ±3 K. (A) Effects of different speciation and dispersal rates on on mean local community species richness. (B) Effects on global metacommunity species richness. (C) Effects on the strength of the latitudinal gradient, represented as the difference between mean equatorial and mean polar richness. Comparison with neutral-metabolic model without temperature niches (see fig. 4.8) reveals only minor differences.

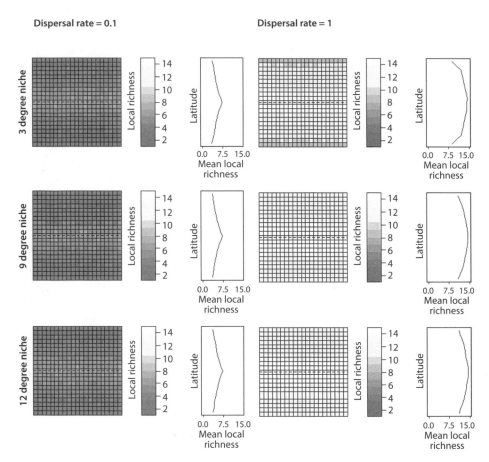

FIGURE 4.12. Effects of changes in thermal niche width and dispersal rate on global gradients in species richness. Shown are metacommunities with the same 30 K temperature gradient and the base model speciation rate $v = 0.01$. At moderate dispersal rate ($m = 0.1$), thermal niches are not affecting gradients of species richness, whereas high dispersal leads to higher richness overall, flatter gradients, and more pronounced latitudinal gradients at reduced niche widths.

originated, as seen in the fossil record (Jablonski et al. 2013). From the macro-ecological perspective, however, and abstracting from species identities, niches appeared to have limited effect on the strength of global richness gradients, except at very high dispersal. Our model results suggest that even moderate dispersal constraints will have a similar effect in constraining most (but certainly not all) species to their original habitat (see fig. 4.9).

With respect to species' ranges, we also found that the distribution patterns that emerged from the model tended to support Rapoport's rule, which states that the latitudinal ranges of plants and animals are generally smaller at lower than at

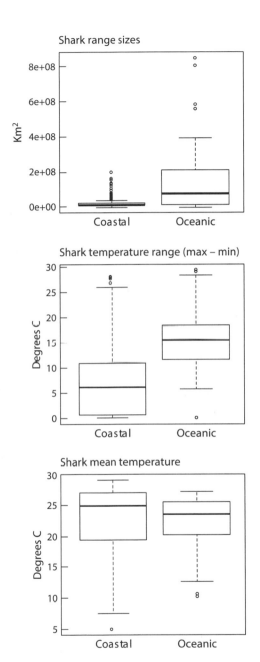

FIGURE 4.13. Realized ranges and thermal niches in coastal versus oceanic sharks. Shown are box-plots of the ranges and temperature requirements of all known shark species, contrasting coastal and oceanic species. Oceanic sharks have both larger ranges and realized thermal niches, but mean temperature preference similar to coastal species. After range data from Lucifora et al. (2011).

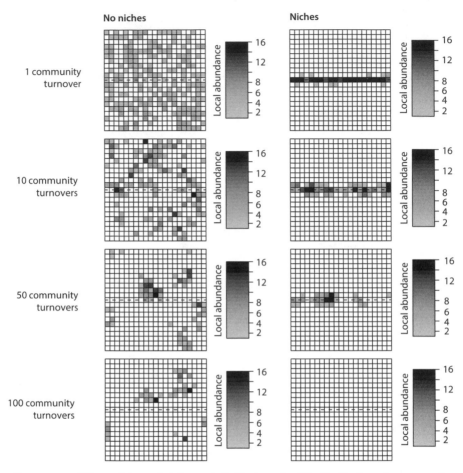

FIGURE 4.14. Effects of thermal niches on individual species. The distribution of two example species is shown over time. It is evident how narrow (3K) temperature niches can constrain the latitudinal spread of a species, its local and global abundance, as well as extinction risk.

higher latitudes (Stevens 1989). Specifically, we saw larger latitudinal ranges for higher latitude species emerge in the model with and without niches when there was also a thermal effect on speciation rate (fig. 4.15). It is noteworthy that this well-documented natural phenomenon could emerge here from a purely neutral model (case with no niches) and did not require previously hypothesized external forces such as seasonality or ice ages as additional explanations. Based on these findings, we propose a new explanation for Rapoport's rule, in that it may emerge from higher relative speciation rates at low latitudes leading to

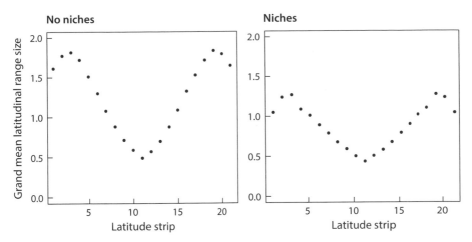

FIGURE 4.15. Rapoport's rule emerging from a neutral model. Average latitudinal range size is plotted versus latitude in a neutral-metabolic model (A) without niches or (B) with ±3K thermal niches. Larger ranges consistently emerge at higher latitudes, possibly as a consequence of lower community richness at those latitudes.

increased numbers of species able to coexist and compete for limiting resources, and hence limiting the range sizes of those same species relative to their high-latitude counterparts.

In conclusion, results emerged from our model that suggest that the effects of temperature on speciation rate and the effects of habitat area and productivity on community size matter most for the equilibrium species richness gradient. These mechanisms correspond to the evolutionary speed and more-individuals hypotheses discussed in chapter 3. Temperature effects on species distribution were also addressed via thermal niches, which had surprisingly little effect on patterns of species richness at the global scale. Niches appeared to have an effect mostly at high dispersal rates, as may be found, for example, in pelagic communities and for widely dispersing species groups such as seabirds, whales, and possibly bacteria. Clearly, niche-width constraints become influential when species have the ability to move about widely and could be more far-ranging were they not limited by thermal constraints. It is under these conditions that we observed effects of niches on the latitudinal gradient, which became steeper when small niches were imposed at high dispersal rates (see fig. 4.12). Overall, however, temperature-driven gradients in speciation rates appeared much more influential than dispersal constraints or niches in determining global richness patterns, at least through the lens of our model. This may partly explain why even species with minimal dispersal constraints can show strong patterns and gradients of biodiversity (Sunagawa et al. 2015).

4.6. DISCUSSION AND COMPARISON WITH OTHER THEORY

The body of theory that is developed here represents only one of many possible perspectives on global biodiversity. At one end of the modeling spectrum, purely statistical models that capture empirical patterns typically perform best in predicting species distribution and diversity (Kreft and Jetz 2007; Tittensor et al. 2010; Belmaker and Jetz 2011; Tisseuil et al. 2013), but provide little mechanistic understanding. At the other end, process-based models provide better understanding, but tend to have poorer predictive capacity (Rohde 1992; Beaugrand et al. 2013; Brown 2014).

Here, we have attempted to merge these perspectives by reviewing statistical associations between observed diversity and environmental drivers (chapters 2 and 3) and hence building a solid empirical basis. We then introduced environmental factors that had gathered the most empirical support into a spatial metacommunity model to assess their effects on the evolution of diversity gradients. Our model is built on, and combines, three of the most prominent ecological theories—namely, neutral, metabolic, and niche theories. The model that emerged from these efforts captures both evolutionary (speciation, extinction) and ecological (dispersal and individual species-environment relationships) dynamics.

Our model may also help to reconcile a long-standing dispute between supporters of niche-based and neutral perspectives. This dispute has been cast as an example of a more general clash between two philosophical perspectives: *realism* and *instrumentalism*, respectively (Wennekes et al. 2012). In the realist's view, a true model will always perform well, precisely because it is true. In contrast, instrumentalists emphasize the predictive value and other uses of a model; the literal truth of the model is less of an issue (Wennekes et al. 2012). Hence, in the niche-neutrality debate, supporters of niche models critique the fact that the key assumption of ecological equivalence in neutral models is clearly unrealistic, while proponents of neutral models are more interested in their predictive capacity (Rosindell et al. 2012). Using art as a metaphor, Wennekes et al. (2012) suggest that neutral theory can provide only broad sketches, while niche theory provides beautiful detail, but leaves much of the canvas blank. They argue that a combined approach would be the most useful way to go about "painting the image," and that this may also be the only way to end up with a reasonable picture. Our theory and model attempt to offer such a combined perspective in the context of a global metacommunity.

In the literature, there is a continuum of purely niche to purely neutral models, for which empirical support falls somewhere in the middle (Gravel et al. 2006; Leibold and McPeek 2006; Adler et al. 2007; Gravel et al. 2011). Some models have attempted to incorporate aspects of both neutral and niche theory, but typically applied at the local community scale and not spatially explicit (Gravel et al. 2006; Leibold and McPeek 2006; Adler et al. 2007). Such models often find that neutral dynamics become more important at high diversity and at larger scales (Chisholm and Pacala 2010), which is relevant with respect to our questions about large-scale biodiversity patterns.

Empirical studies have likewise found evidence for both neutral processes (ecological drift) and environmental filtering via niches to explain community structure. For example, a large-scale study conducted along a meridional transect of the Atlantic (50 degrees North to 50 degrees South) suggested that phytoplankton communities were slightly more determined by niche segregation (24%), than by dispersal limitation and ecological drift (17%). Notably, in 60% of the surveyed communities, the assumption of neutrality in species' abundance distributions could not be rejected (Chust et al. 2012).

Plankton communities also served as model systems in a previous attempt to reconstruct global diversity patterns from first principles (Beaugrand et al. 2013). In that study, hypothetical "pseudospecies" were randomly assembled along a latitudinal temperature gradient. Purely by chance, more species assembled toward the center of the temperature domain, in a niche analogy to the spatial mid-domain effect (see also chapter 3). Accounting for assumed loss of species at higher latitudes owing to long-term climatic variability (such as ice ages), a gradient that is broadly representative of those observed in many pelagic groups could be constructed (Beaugrand et al. 2013; Beaugrand 2014). The question that emerges from this work is whether niches are indeed necessary to explain observed diversity patterns. We observed that their effect is mostly small in our model, because most species tended to have small latitudinal ranges regardless of assumed niche width constraints (see fig. 4.9). In our fully neutral model, it appeared that niches were not a major driver of species distributions, except when there was no other limitation to dispersal. In other words, temperature niches appeared here to be more a consequence rather than a root cause of the eco-evolutionary processes that shape global biodiversity patterns. We see these results as a starting point to dive deeper and to further explore the rich variety of diversity patterns found in different realms—namely, terrestrial, coastal, pelagic, and deep-sea environments. This is the topic of our next chapter, where we attempt to confront our theory with the existing data on species richness patterns in the ocean and on land.

CHAPTER FIVE

Predicting Global Biodiversity
Patterns from Theory

In the previous chapter, we developed a global theory of biodiversity incorporating gradients in ambient temperature and habitat area or productivity. We showed that a metacommunity model implementation of our theory can reproduce first-order patterns of declining species richness from the tropics to the poles in an idealized cylindrical ocean. Low-dispersal scenarios were more likely to reproduce steep latitudinal declines with a tropical peak, such as those seen in coastal and terrestrial communities, whereas high metacommunity dispersal rates produced flatter gradients, such as those observed in many pelagic communities. The relaxation of our neutrality assumption and the inclusion of thermal niches had little effect on latitudinal richness patterns except at very high dispersal rates.

In this chapter, we test our theory in a more realistic setting by fitting the neutral-metabolic metacommunity model (chapter 4) to a global equal-area grid (chapter 2) with a more realistic spatial structure and empirically observed gradients in temperature, habitat, and productivity variables that were identified as potential drivers in chapter 3. The rationale here is to explore whether the communities that evolve in a simple theoretical model can reproduce observed patterns of species richness in the real world, and reconcile the contrasting patterns seen in coastal, pelagic, deep-sea, and terrestrial habitats. Ultimately, we aim to assess the extent to which our theory and its model implementation can capture the processes that structure spatial patterns of biodiversity at the global scale.

Our core assumption is that the fundamental ecological and evolutionary mechanisms by which biodiversity is generated should be consistent across realms and taxa, but with the unique patterns in each resulting from different environmental constraints in different habitats. In chapter 3, we showed empirically that temperature is the primary environmental variable that correlates with most global richness patterns, with the exception of some endotherms and cold-water specialists in polar waters and the deep sea. Other important environmental predictors capture habitat area and productivity; however, they differed among realms and species groups. They included (1) coastline length in coastal species,

(2) oceanographic fronts for pelagic groups, (3) export production for deep-sea taxa, and (4) water availability on land, all of which influence habitat availability or productivity, and hence community size. Empirically, we have found that these habitat features, in conjunction with temperature, correlate with and help to predict observed patterns of species richness (chapter 3). Mechanistically, this is likely the case because greater habitat area or productivity allows for higher numbers of individuals to coexist in a given habitat, and hence larger community size, larger populations sizes, and reduced extinction rates (Wright 1983; Gaston 2000). Empirical observations strongly support the notion that a larger area of shoreline supports more individuals (Abele and Patton 1976; Connor et al. 2000), frontal areas sustain higher densities of pelagic animals (Haney 1986; Olson et al. 1994), areas of high export production support larger communities in the deep sea (Haedrich and Rowe 1977; Rex and Etter 2010), and wetter areas allow for more biomass to develop on land (Silvertown et al. 1994; Kreft and Jetz 2007), all else being equal. By incorporating these habitat features into a metacommunity model with realistic spatial structure, and scaling them to the number of individuals in our model communities, we attempt to reproduce broad patterns of biodiversity observed in these four major habitats. Much like in the previous chapter, we do try to keep the model framework as simple and tractable as possible, or "instrumental" rather than "realistic" *sensu* (Wennekes et al. 2012). Time steps, community sizes, and speciation rates as implemented in our model, for example, remain abstractions and are not directly comparable to the real world. We also do not employ a formal statistical model selection framework to optimize model fit to data. Rather, to aid in our understanding, we explore a set of base parameters that match the data reasonably well, and then we allow these parameters to vary such that we can explore the sensitivity of model predictions and their fit to observed data.

5.1. FITTING THEORETICAL PREDICTIONS TO EMPIRICAL DATA

To facilitate the comparison of theoretical predictions with empirical richness data, we scaled our global metacommunity model to the same 880 × 880 km equal-area grid used in chapters 2 and 3. Across this global grid, 325 cells contained coastal areas, 526 were pelagic, 500 deep sea (<2000 m), and 205 on land. We included a cell, for example as coastal or deep sea, if it had at least 10% of its area covered by that respective realm. Every cell then equated to a local community. The basic local community size was assumed to be $J = 16$ individuals, as in our base model, but this number was allowed to increase in relation to changes in habitat area or productivity (see the following). Metacommunity size

J_M was variable, and a product of the number of cells in each realm (between 205 and 526) and the number of individuals per community (between 16 and 240). As in chapter 4, we assumed that dispersal occurs only between adjacent cells (local communities), and that cells outside that particular realm would block dispersal. Hence, terrestrial communities could not disperse across marine cells, coastal communities could not disperse across terrestrial or noncoastal marine cells, and so forth. The Moore neighborhood of each local community therefore varied based on the geography of the grid cell. Further, we assumed that there was no dispersal through the Suez and Panama canals, which could act as dispersal pathways between the Atlantic and Indo-Pacific. As these pathways were opened relatively recently, we surmised that they may not yet have influenced global richness patterns in a major way. When we tested the sensitivity of predicted richness patterns to this particular assumption (by opening or closing canals), we found little effect (results not shown). We explored the land, coastal, pelagic, and deep-sea communities separately in our analysis by running ensembles of simulations for each realm.

The effects of temperature on turnover and speciation rates were parameterized using metabolic theory, and were exactly analogous to chapter 4, but driven by the observed global (surface) temperature field in both latitude and longitude, rather than the hypothetical 0 to 30 degree gradient that was assumed in chapter 4. Global sea and land surface temperature data were taken from published sources as documented in chapter 3.

While the relationships between temperature, metabolic rate, community turnover, and speciation rate are based on exact mathematical expressions, there is no analogous expression that enables us to generally parameterize a relationship between empirical measures of habitat area or productivity and the number of individuals in a local community. While an approximately linear or log-linear relationship is empirically founded, the slope of that relationship may be variable and has so far been assessed only over small spatial scales (Connor et al. 2000). Therefore, we scaled the effects of habitat area, productivity, and moisture (on land) in a linear fashion to the number of individuals in a local community, analogous to chapter 4. This reflects the empirically observed linear individuals-area relationship for whole faunas (Connor et al. 2000).The relationship between the number of individuals and habitat area may alternatively be derived theoretically through consideration of two other well-documented patterns: the species-area relationship and the species-individuals relationship (Peet 1974). These two relationships, when combined, give rise to an individuals-area relationship. However, given the wide range of observed slopes values for species-area and species-individuals relationships (Hillebrand et al. 2001; Drakare et al. 2006), the slope of the relationship between individuals and area will also vary widely. We further note that these relationships may change at large scales, as for example in the species-area slope, which was shown to be scale dependent (Crawley and Harral 2001).

We conclude that there is both theoretical and empirical justification for a linear relationship between the number of individuals and habitat area, but that the slope value of that relationship is unknown. This is even more salient for measures of productivity, which likewise should increase the density of individuals per unit area, possibly in a linear fashion, but with unknown slope. One study suggests that at global scales, the relationship between species (as opposed to individuals) and energy is steeper than that between species and area (Hurlbert and Jetz 2010). We thus explored a range of possible scaling relationships, by varying the slope of the linear function relating empirical habitat and productivity variables (coastline length, frontal zones, export productivity, or moisture respectively) to the number of individuals in a local model community, treating this slope as a parameter to be estimated. We normalized the smallest value of each empirical habitat and productivity variable to one, with linear scaling to higher values, and estimated the sensitivity of model predictions to different slope values. We explored a wide range of scaling relationships, but generally found that a two- to fivefold increase in community size with increasing habitat area or productivity provides the best fit to data (see figs. 5.1 to 5.7). Once we identified an approximate optimal scaling of community sizes, we explored the sensitivity of model predictions to varying dispersal rate ($1 \leq m \leq 0.001$), and speciation rate ($0.1 \leq v \leq 0.0001$), as in chapter 4. All parameter combinations were run 100 times in coalescence mode. We took the mean community values across the 100 replicate runs to average across the stochastic variation inherent in these simulations.

In the following, we separate our model prediction exercise between coastal, pelagic, deep-sea, and land species, because these major environmental realms each showed distinct patterns of global biodiversity (chapter 2). We aim to explore similarities and differences between relative richness patterns (normalized to maximum value), and do not attempt to estimate absolute values of species richness. We further explore model predictions for endotherms versus ectotherms, because their metabolic rates may have different sensitivity to ambient temperature. Importantly, we fit only to observed richness data, and do not use any environmentally extrapolated data points, to prevent circularity of reasoning. Finally, we assess under which conditions the inclusion of temperature niches improves model predictions of realized biodiversity patterns.

5.1.1. Coastal Biodiversity

First, we used the model to predict a generalized biodiversity pattern of coastal ectotherm species, which was derived as a normalized composite of observed coral, bivalve, mangrove, seagrass, ophiuroid, cone snail, cephalopod, coastal fish, and coastal shark richness (fig. 5.1; results for single taxa can be seen in table 5.1). We did not include the pinnipeds, owing to their endothermy and specialized cold-adaptation, and made predictions for coastal cells only. When we ran our metacommunity model on this coastal grid ($v = 0.001$, $m = 0.1$), a null-model effect was examined first, which was a base-parameter model run that included the observed spatial structure of present ocean basins but ignored the effects of temperature or community size (that is, a basic neutral metacommunity model). Such a null-model correlated positively, but weakly, with observed richness ($r = 0.37$, t-test $P < 0.001$). This fit is entirely due to the spatial structure of today's oceans, including connectivity and continental dispersal barriers as well as edge effects in polar regions. The relative importance of other processes can be assessed against such a null-model.

When we added the effect of observed surface temperature on speciation rates to this model, the predicted global richness pattern became much more similar to the empirical data ($r = 0.70$, $P < 0.0001$), except slightly overpredicting diversity in the central and eastern Pacific.

When including the effects of both temperature and coastline length on community size in our model, the fit with observed data improved ($r = 0.77$, $P < 0.0001$; fig. 5.2A), but not dramatically. Scaling the effects of coastline length to community size by a factor of 3 over the baseline model provided the optimal fit (fig. 5.2B). Everything else being equal, longer coastlines will harbor more individuals, influencing community size, but also tend to display larger habitat complexity due to the inclusion of bays and islands, for example. In contrast, inclusion of net primary productivity (NPP) as an alternative variable that influences community size provided a weaker prediction than the effect of coastline length (fig. 5.2B). Clearly, the majority of the observed pattern related to temperature variation and its effects on evolutionary rates, with habitat area (and possibly habitat complexity) playing an additional role.

While our theoretical model captured the main global features observed for coastal ectotherm richness (see fig. 5.1A,B), it did not predict the steepness of observed gradients in normalized richness as well. Most areas were slightly overpredicted, while species richness across the Indonesian-Australian Archipelago hotspot was underpredicted by our model (see fig. 5.1C). We speculate that high habitat and trophic complexity in that hotspot may further amplify diversity in a way that is not captured by a simple measure of coastline length. Species interactions and coevolution, as well as speciation by reproductive isolation in dynamic and structurally complex coral reef habitats may provide additional mechanisms for diversification not captured by our model (Bellwood et al. 2012). This is the "diversity begets diversity" argument, which does not explain the existence of a

TABLE 5.1. Model Fit to Global Biodiversity Data

Habitat	Species group	Latitudinal peak	Accuracy	Longitudinal peak	Accuracy	r	P
Coastal	All ectotherms	Tropical	Yes	East Asia	Yes	0.77	< 0.0001
Coastal	Stony corals	Tropical	Yes	East Asia	Yes	0.61	< 0.0001
Coastal	Brittle stars	Tropical	Yes	East Asia	Yes	0.70	< 0.0001
Coastal	Bivalves	Tropical	Yes	East Asia	Yes	0.71	< 0.0001
Coastal	Cone snails	Tropical	Yes	East Asia	Yes	0.63	< 0.0001
Coastal	Cephalopods*	Subtropical	No	East Asia	Yes	0.57	< 0.0001
Coastal	Mangroves	Tropical	Yes	East Asia	Yes	0.69	< 0.0001
Coastal	Seagrasses	Subtropical	No	East Asia	Yes	0.48	< 0.0001
Coastal	Fish	Tropical	Yes	East Asia	Yes	0.70	< 0.0001
Coastal	Sharks	Subtropical	No	East Asia	Yes	0.36	< 0.0001
Coastal	Pinnipeds	Polar	No	Various	No	−0.39	< 0.0001
Pelagic	All ectotherms	Subtropical	Yes	Various	Yes	0.87	< 0.0001
Pelagic	Foraminifera	Subtropical	No	Various	Yes	0.85	< 0.0001
Pelagic	Euphausiids	Subtropical	No	Various	Yes	0.78	< 0.0001
Pelagic	Squids*	Temperate	No	Various	Yes	0.20	< 0.0001
Pelagic	Tuna and billfish	Tropical-subtropical	Yes	Various	Yes	0.64	< 0.0001
Pelagic	Sharks	Subtropical	No	East Asia	Yes	0.59	< 0.0001
Pelagic	Cetaceans	Temperate	No	Various	Yes	0.70	< 0.0001
Pelagic	Seabirds	Temperate-polar	No	Various	Yes	−0.14	< 0.01
Deep sea	All ectotherms	Temperate	Yes	Various	Yes	0.24	< 0.01
Deep sea	Brittle stars	Temperate	Yes	Various	Yes	0.24	< 0.01
Land	All ectotherms	Tropical	Yes	Various	Yes	0.70	< 0.0001
Land	Vascular plants	Tropical	Yes	Various	Yes	0.77	< 0.0001
Land	Amphibians	Tropical	Yes	South America	No	0.64	< 0.0001
Land	Reptiles*	Subtropical	No	Central America, Africa, Asia	No	0.32	< 0.0001
Land	Birds	Tropical	Yes	South America, Asia	No	0.59	< 0.0001
Land	Mammals	Tropical	Yes	South America, Africa, Asia	Yes	0.62	< 0.0001
Success rate			60%		88%		92%

Note: Shown are the summary results of fitting neutral-metabolic metacommunity model runs to empirical data for aggregate groups (see also figs. 5.1 to 5.8) and fits of the same model for individual taxa. The strength and significance of the correlation is indicated by correlation coefficient *r* and associated *P*-value. Accuracy of predicted versus observed latitudinal and longitudinal peaks in species richness is noted. Limited global sampling or limited species coverage is marked with an asterisk (*).

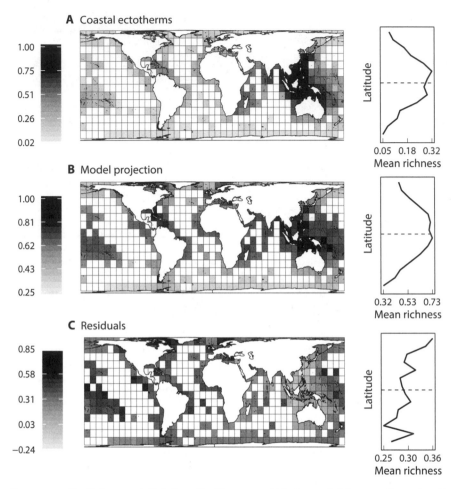

FIGURE 5.1. Predicting coastal biodiversity. Shown are (A) observed richness patterns for all coastal ectotherm species combined, and (B) predicted global richness patterns derived from a neutral-metabolic metacommunity model ($m = 0.1$, $v = 0.01$) involving the effects of sea surface temperature on evolutionary speed and the effects of coastline length on community size. (C) Residual variation plot displays spatial structure in the fit between theoretical model predictions and empirical data. Note that only cells with both model prediction and empirical data have been plotted.

gradient, but may explain why some of the gradients are steeper than expected by metabolic theory alone (Brown 2014).

When exploring the sensitivity of theoretical model predictions to changes in speciation and dispersal parameters (see fig. 5.2C), we found moderate sensitivity to variation in speciation rates, but low sensitivity to changes in dispersal rates, except at very low speciation, which produced a flat diversity pattern that did not

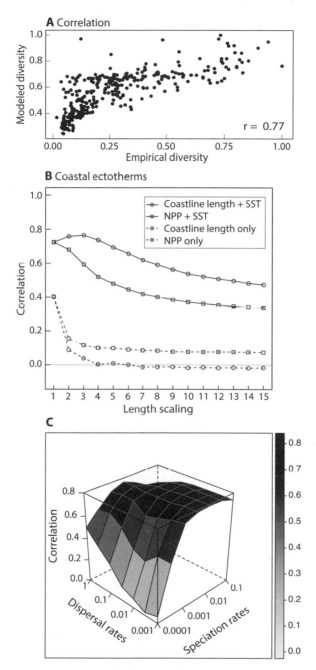

FIGURE 5.2. Sensitivity of predictions for coastal biodiversity. (A) The best model fit is derived by the inclusion of metabolic effects of sea surface temperature (SST) on evolutionary speed and the effects of coastline length on community size. (B) Sensitivity of model predictions to different correlates of community size (coastline length or net primary productivity) is shown at a range of scaling factors. (C) Sensitivity of model results to variation in speciation and dispersal rates.

fit well to observed data for any species group. Moderate speciation and dispersal rates ($m = 0.1$, $v = 0.001$), as used in most of our simulations, produced the best fit to the observed data, with low sensitivity to modest changes in these parameters (see fig. 5.2C).

5.1.2. Pelagic Biodiversity

Pelagic species were combined to derive a normalized average richness pattern for tuna and billfish, pelagic sharks, squids, euphausiids, and foraminifera (fig. 5.3; results for single taxa can be seen in table 5.1), but excluding cetaceans due to their endothermy. Similar to the coastal species, the corresponding pelagic richness pattern was predicted surprisingly well by the effects of temperature alone ($r = 0.77$, $P < 0.0001$), an effect that was improved by including the effect of pelagic frontal habitats on community size ($r = 0.87$, $P < 0.0001$; fig. 5.4A). Again, scaling the effects of habitat to include a twofold increase in community size provided the best fit to empirical data, inclusion of NPP gave a slightly poorer fit to the data, and exclusion of sea surface temperature (SST) effects provided a poor fit overall (fig. 5.4B). It is interesting to note that the primary mechanisms—that is, a large effect of temperature on evolutionary speed and smaller effects of habitat on community size—appear similar between coastal and pelagic ectotherms despite contrasting global richness patterns (see figs. 5.1 and 5.3). In contrast to the coastal model, however, the pelagic model predicted species richness most accurately in tropical regions, while underpredicting some temperate and overpredicting some polar regions, although not dramatically (fig. 5.3C). We note that the fit from a process-based neutral-metabolic model is almost as good as that from a purely statistical model that includes more free parameters and explanatory variables (Tittensor et al. 2010). Predictions were sensitive to low speciation rates and high dispersal rates, but stable over much of the explored parameter space (fig. 5.4C).

5.1.3. Deep-Sea Biodiversity

Deep-sea ophiuroids represented the single deep-sea taxon with appropriate global sampling. Their global richness pattern was predicted by the effects of export productivity, as scaled to community size in our theoretical model (fig. 5.5), but with a modest overall fit ($r = 0.24$, $P < 0.01$; fig. 5.6A). Including the effects of deep-water temperature made the prediction slightly worse (fig. 5.6B), and did not give much contrast, as temperatures are uniformly cold (2–4°C) below 2000 m depth.

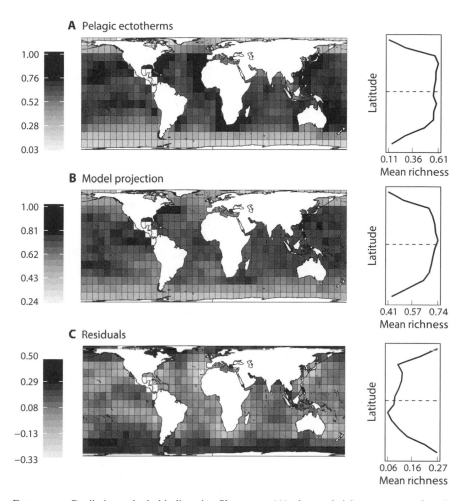

FIGURE 5.3. Predicting pelagic biodiversity. Shown are (A) observed richness patterns for all pelagic ectotherm species combined, and (B) predicted global richness patterns derived from a neutral-metabolic metacommunity model ($m = 0.1$, $v = 0.01$) involving the effects of sea surface temperature (SST) on evolutionary speed and pelagic frontal habitats (measured as SST slope) on community size. (C) Residual variation plot displays spatial structure in the fit between theoretical model predictions and empirical data. Note that only cells with both model prediction and empirical data have been plotted.

Model predictions of observed richness tended to improve at moderate dispersal and low speciation rates, and were somewhat weaker at low dispersal and high speciation rate (fig. 5.6C). There was little pattern in the residuals, suggesting that an average pattern was captured by our theoretical model, but that there is significant spatially random variation that cannot be readily explained.

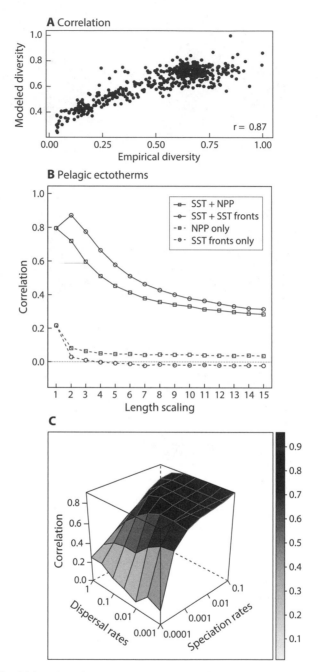

FIGURE 5.4. Sensitivity of predictions for pelagic biodiversity. (A) The best model fit is derived by the inclusion of metabolic effects of sea surface temperature (SST) on evolutionary speed and pelagic frontal habitats (measured as SST slope) on community size. (B) Sensitivity of model predictions to different correlates of community size (SST slope or net primary productivity) at a range of scaling factors. (C) Sensitivity of model results to variation in speciation and dispersal rates.

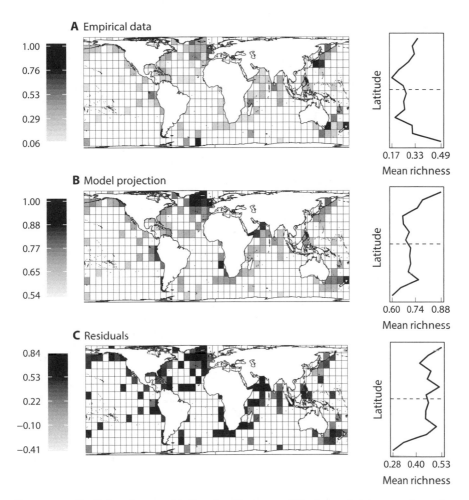

FIGURE 5.5. Predicting deep-sea biodiversity. Shown are (A) observed richness patterns for deep-sea ophiuroids, and (B) predicted global richness patterns derived from a neutral-metabolic metacommunity model ($m = 0.1$, $v = 0.01$) involving the effects of export productivity on community size. (C) Residual variation plot displays spatial structure in the fit between theoretical model predictions and empirical data. Note that only cells with both model prediction and empirical data have been plotted.

5.1.6. Terrestrial Biodiversity

Normalized richness patterns for vascular plants, amphibians, and reptiles were combined to examine model fit for terrestrial ectotherm species (fig. 5.7; results for single taxa can be seen in table 5.1). Mammals and birds were excluded due to their endothermy and are treated separately later. Adapting our neutral-metabolic

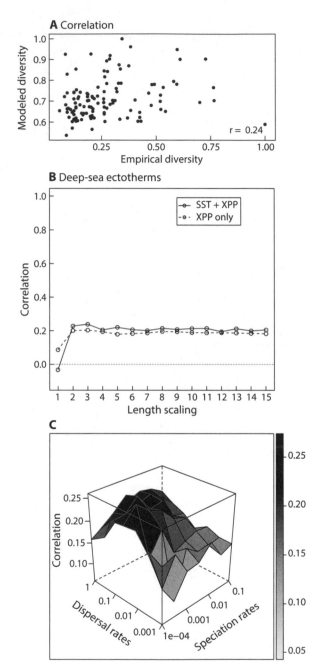

FIGURE 5.6. Sensitivity of predictions for deep-sea biodiversity. (A) The best model fit is derived by the inclusion of export production (XPP) on community size. (B) Sensitivity of model predictions to the inclusion of temperature or XPP at a range of scaling factors. (C) Sensitivity of model results to variation in speciation and dispersal rates.

metacommunity model to the land required including the effects of moisture, expressed as the average number of wet days per year. Community size was scaled to this variable, predicting that wetter places support more individuals and larger communities of land plants and other ectotherms, and hence more species, all else being equal. We found that together with the effects of temperature, this provided a surprisingly good prediction of the observed richness pattern at a global scale (figs. 5.7 and 5.8A). Some species-rich hotspots in the Andes or in South Asia were underpredicted (fig. 5.7C), possibly due to the added effects of elevated habitat complexity and elevation range found there (Kreft and Jetz 2007; Belmaker and Jetz 2011; Weigelt et al. 2016). Despite these local residuals, however, the overall predicted pattern fit the observed richness data surprisingly well ($r = 0.70$, $P < 0.0001$), with an even better fit achieved for plants only ($r = 0.77$, $P < 0.0001$; see table 5.1), which represent the most well-sampled taxon, and maybe most strongly relate to the effects of moisture (wet days), as implemented in this model. The improvement of model fit by including the number of wet days was substantial over the temperature-only model (fig. 5.8B), and maintained across a range of scaling relationships. Predicted patterns and their fit to empirical data were not very sensitive to variation in parameters, with the exception of low speciation rates, which provide a low fit due to low equilibrium richness overall (fig. 5.8C).

5.2. ECTOTHERMS VERSUS ENDOTHERMS

As metabolic theory is thought to apply specifically to ectotherms, we analyzed endothermic mammals and birds separately. Somewhat surprisingly, we found that the effects of surface temperature as implemented in the model also needed to be invoked to explain endotherm biodiversity patterns. Cetacean richness, for example, was quite well predicted by the effects of SST as implemented in our theoretical model ($r = 0.52$, $P < 0.001$). This prediction was improved upon by adding the effects of NPP or fronts on community size, with fronts being the better predictor of community size (fig. 5.9A). When only the effects of fronts or NPP were implemented, these models predicted the observed data slightly better than the SST-only model. However, the best model included both SST and fronts, similar to the pelagic ectotherms, but with a weaker fit to the observed data ($r = 0.70$, $P < 0.0001$). Model residuals were clustered in coastal areas and in the Southern Ocean, where richness was underpredicted by our theoretical model, relative to observed data (fig. 5.9A).

The cetaceans' mammalian cousins on land provided an interesting point of comparison (see fig. 5.9B). Again, we were surprised to find that the effects of

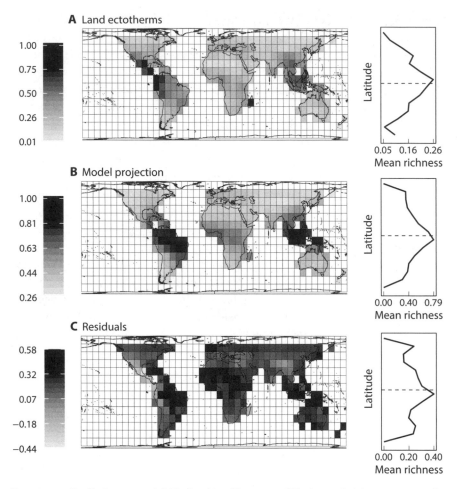

FIGURE 5.7. Predicting terrestrial biodiversity. Shown are (A) observed richness patterns for land ectotherms (land plants, amphibians, reptiles), and (B) predicted global richness patterns derived from a neutral-metabolic metacommunity model ($m = 0.1$, $v = 0.01$) involving the effects of temperature on evolutionary speed and the effects of the number of wet days on community size. (C) Residual variation plot displays spatial structure in the fit between theoretical model predictions and empirical data. Note that only cells with both model prediction and empirical data have been plotted.

temperature on metabolic rate as implemented in our theoretical model provided a reasonable prediction of the observed data. Community size was scaled to the number of wet days, or alternatively to the rate of NPP, which provided a marginally better fit (not shown). Interestingly, most negative residuals were found in mountainous regions, where the model consistently underpredicted species

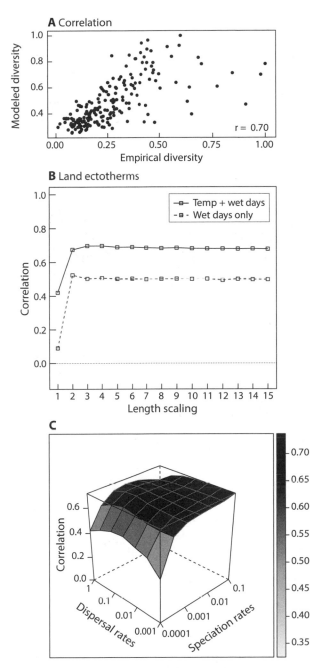

FIGURE 5.8. Sensitivity of predictions for terrestrial biodiversity. (A) The best model fit is derived by the inclusion of metabolic effects of surface temperature on evolutionary speed and the effects of the number of wet days on community size. (B) Sensitivity of model predictions to the inclusion of wet days and temperature at a range of scaling factors (C) Sensitivity of model results to variation in speciation and dispersal rates.

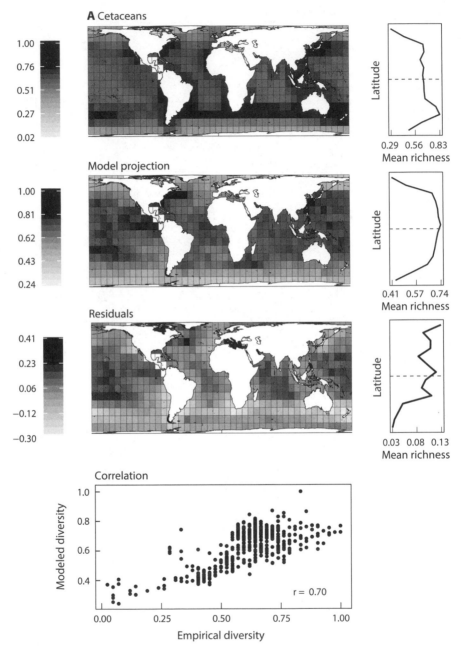

FIGURE 5.9. Predicting biodiversity of endotherms. Shown are observed and predicted global richness patterns for (A) marine mammals (cetacean), and (B) land mammals from a neutral-metabolic metacommunity model ($m = 0.1$, $v = 0.01$), including the effects of surface temperature and wet days on land, and sea surface temperature and net primary productivity in the

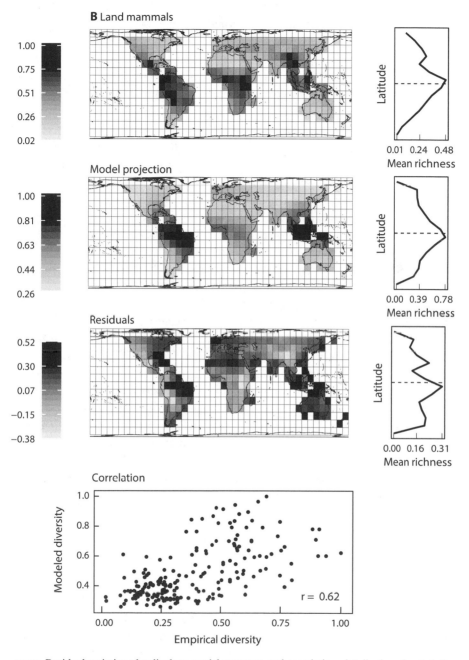

B Land mammals

Model projection

Residuals

Correlation

ocean. Residual variation plot displays spatial structure, and correlation plot displays the statistical fit between theoretical model predictions and empirical data. Note that only cells with both model prediction and empirical data have been plotted.

richness. As for land ectotherms, we surmise that the effects of topographic heterogeneity may be important here, although we cannot at present capture this effect in our model. The overall fit relative to observed data ($r = 0.62$, $P < 0.0001$) was somewhat weaker than for land ectotherms, and sensitivity to change in the dispersal parameter was more pronounced, with low predictive power at both the high and low end of the scale for dispersal rate (not shown).

The other endotherm taxon examined was birds, which were almost equally well predicted using the same model as employed for mammals on land ($r = 0.59$, $P < 0.0001$). However, pelagic seabirds (Order: Procellariformes) could not be predicted well at all ($r = -0.14$, $P < 0.01$; see table 5.1) owing to a heavily skewed distribution of seabird diversity toward the Southern Ocean (see fig. 2.6). Possibly, the unique dependence of this taxon on wind energy may contribute to an outlying pattern of diversity, as suggested by Davies et al. (2010). Similarly, the pinnipeds with their polar distribution were not captured by our metacommunity model in the same way that cetaceans were ($r = -0.39$, $P < 0.001$; see table 5.1), likely owing to their unique circumpolar distribution.

In conclusion, our theory performs reasonably well for several larger endotherm groups comprising thousands of species (land mammals and birds, cetaceans) but poorly for smaller, more specialized marine groups found at higher latitudes (procellariform seabirds and pinnipeds). The fact that major species groups are captured reasonably well by the theory may point toward an unexplained mechanism that links temperature to evolutionary rates in endotherms, for which we might otherwise not expect such a pattern. Alternatively, or in addition, there may be biotic interactions with ectotherms such as competition, predation, or mutualism that affect the biogeography of endotherms, resulting in an apparent relationship with temperature. Such interactions might also in part explain the specialized distributions of endotherm taxa that find a competitive niche in colder, highly productive environments.

5.3. INCLUDING NICHES

In chapter 4, we hypothesized that global richness patterns may in part relate to species temperature tolerances and associated thermal niches. The effects of niches in constraining realized range size and dispersal, however, was very modest, except in high-dispersal scenarios ($m = 1$), where smaller niches constrained range sizes and led to a steepening of latitudinal gradients, in comparison to non-niche scenarios (see figs. 4.9 to 4.12). We further hypothesized that such a high dispersal scenario may approximate pelagic communities, where average range sizes tends to be larger (see fig. 4.13) and there are few constraints to dispersal, other than unsuitable temperature regimes (Beaugrand et al. 2013).

Here, we evaluate the effects of assumed temperature niches in our expanded neutral-metabolic-niche model. We began with a base case of moderate dispersal and speciation rates ($m = 0.1$, $v = 0.01$) and used the best-fitting model for pelagic biodiversity, including the metabolic effects of SST on turnover and speciation as well as the effects of frontal habitats on community size (see also see fig. 5.3). When we introduced nonevolving or evolving thermal niches of various width (from $w = \pm3$ K to $w = \pm12$ K) into this base model, the predicted patterns of pelagic species richness were practically indistinguishable from the case without niches, and the statistical fit to observed data was very similar ($r = 0.81$, $P < 0.0001$). In order to check this negative result in another environment, and for less mobile species, we applied the same procedure to land plants, and again got identical results to the base case without assumed niches, with identical fit to observed data ($r = 0.77$, $P < 0.0001$).

Next, we expanded to a scenario of unconstrained dispersal ($m = 1$) in pelagic communities, where such high dispersal rates may be more plausible. In this case, the ability of the neutral-metabolic base model to capture observed richness patterns for pelagic ectotherms was poor ($r = 0.54$), largely because the high dispersal rate equalized global richness and led to a flat pattern overall (fig. 5.10A). When introducing narrow niches ($w = \pm3$ K) into this model, these formed a barrier to dispersal and partially restored a gradient in species richness (fig. 5.10B), and henceforth provided an improved fit to observed data ($r = 0.81$). Yet these predictions were not as strong as for the non-niche case at moderate dispersal (see figs. 5.3 and 5.4), which also provided a simpler model with fewer assumed parameters. These results mirror and support our previous conclusions from chapter 4 that niches have little effect on equilibrium species richness patterns in our model, except for the limiting case of unconstrained dispersal. Unconstrained dispersal, however, did not provide a better prediction of the observed data, even for highly mobile pelagic communities.

5.4. SYNTHESIS

First, we shall pause to emphasize the difference between our theoretical model and other approaches to predicting global patterns of biodiversity. While the common approach is to statistically predict observed patterns from environmental correlates (chapter 3 and references therein), our approach here is very different. Based on our combined neutral-metabolic theory, we have simulated

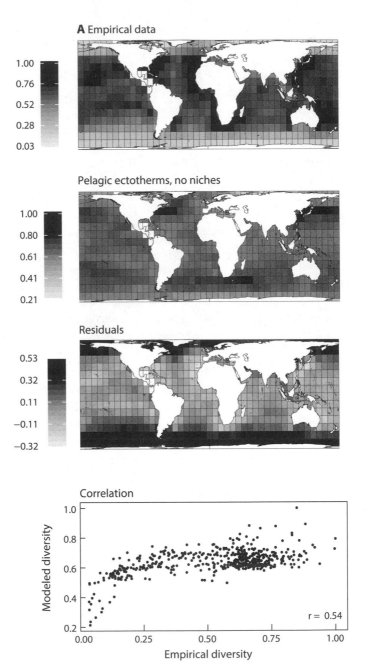

FIGURE 5.10. Effects of thermal niches on model predictions at high dispersal rates. Shown are predicted global richness patterns for pelagic ectotherm species derived from a neutral-metabolic metacommunity model, and compared to observed data. The effects of sea surface temperature (SST) and frontal habitat are included, as in fig. 5.3, but dispersal rate is assumed to be elevated ten-fold ($m = 1$, $v = 0.01$). (A) Empirical composite pattern of species richness for

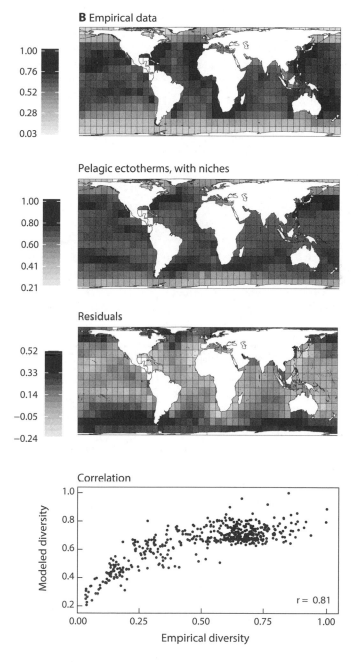

pelagic ectotherms and model predictions, residuals and correlation plots assuming no niches; and (B) assuming ±3 K thermal niches. Inclusion of niches improved the predictions under assumed high dispersal rate, but did not produce as good a fit as the non-niche model at moderate dispersal rate (see figs. 5.3 and 5.4).

ecological processes (disturbance, dispersal) and evolutionary processes (specia-tion, extinction), allowing local and global community structure to evolve from these mechanisms without predetermination. The theoretical model we used here for prediction is very simple, including a maximum of four free parameters (scaling of local community size J as a function of habitat area or productivity), as well as speciation rate v, dispersal rate m, and thermal niche width w. This contrasts markedly with ecological models that typically feature dozens of free parameters (Pauly et al. 2000; Fulton et al. 2005; Harfoot et al. 2014) and with statistical models, in which a relationship between an environmental predictor and observed data is estimated without necessarily shedding light on a mecha-nism (Gotelli et al. 2009; Tittensor et al. 2010).

It appears remarkable that at most two mechanistic processes (metabolic scal-ing of speciation rate with temperature and scaling of community size with habitat area or productivity) need to be invoked to predict in some detail observed global patterns of biodiversity, generating fits to empirical data that are within the general range of statistical nonmechanistic models (see table 5.1). Even more surprising is the fact that these same two mechanisms in our process-based models fit across biomes, reconciling previously disparate patterns of species richness on land and in the sea, for highly mobile and sessile species, and to some degree for endo-therms and ectotherms. This provides support for the idea that the processes that we discuss here are indeed of fundamental importance for the evolution of life forms and biodiversity at large scales. Results further suggest that a reasonably simple theoretical model can aid us in understanding and distinguishing between the potential mechanisms—some of which may be indistinguishable through sta-tistical fits—operating in real-world communities and shaping global biodiversity.

When we compared theoretical model predictions across realms and species groups, we found that the fit to observed data differed (see table 5.1). The best predictions were achieved for aggregate ectotherm groups (shown as subheadings in table 5.1). The models for pelagic ectotherm biodiversity fit best ($r = 0.87$), followed by coastal ($r = 0.77$), and land species groups ($r = 0.70$), whereas model predictions for the sole taxonomic group in the deep sea fit observed data more poorly ($r = 0.24$; see table 5.1). Empirically, the deep sea is the most poorly sam-pled, but it also provides a special case from a modeling perspective because of a flat and uniformly cold temperature profile. Hence, it might be unsurprising that our neutral-metabolic model does not perform as well here.

Individual taxa that were best predicted by our model included some coastal and pelagic invertebrates as well as land plants. It was notable that endotherm taxa generally did not fit as well as their ectotherm counterparts in the same habitat, with negative predictions recorded for pinnipeds and seabirds. This likely indi-cates that the metabolic theory model works best for ectotherms, and additional processes may need to be invoked to understand the distribution of endotherm species at a global scale. There was also a contrast in the sensitivity of model predictions to changes in dispersal and speciation parameters. Ectotherms were most sensitive to changes in speciation rates, and generally fit better when assuming moderate to high speciation rates. Model predictions for endotherms, in contrast, appeared more sensitive to variation in dispersal rates. Our standard scenario of moderate speciation and dispersal rates ($m = 0.1$, $v = 0.01$), however, fit almost equally well across groups. In addition to endotherm species groups, taxa with limited geographic or species coverage were also not predicted as well as others (see table 5.1).

As seen in chapter 4, model predictions here were less sensitive to the inclusion of thermal niches, except at very high dispersal rates. Such high-dispersal scenarios, however, did provide a lesser fit to observed data than moderate dispersal rates. This leads us to conclude that a base case without niches provides a more parsimonious explanation, as fewer parameters are invoked, and the predictive capacity is the same as for an extended case that includes niches. We reiterate here that temperature niches are undoubtedly important in driving species composition and turnover, but appear less influential on global biodiversity patterns, at least through the lens of our model. Possibly, niches are better understood as an emergent outcome of mechanisms generating biodiversity, rather than as a mechanism in themselves.

In summary, our theoretical model does surprisingly well not just in delineating broad latitudinal pattern in a theoretical ocean (chapter 4) but also in capturing more complex patterns of observed species richness in the real world (see table 5.1). Despite significant variability among habitats and individual taxa, 60% of taxa were successfully predicted in terms of their latitudinal peak and 88% in terms of their general longitudinal pattern of diversity, and 92% showed a positive and significant ($P < 0.01$) fit to the observed data overall (longitude × latitude full spatial pattern) (see table 5.1).

The model is simple and abstract, and most certainly misses important processes, particularly in diversity hotspots, where it often underpredicts observed richness. Here, additional environmental variables such as habitat, topographical, or vegetation complexity may play an important role (Kreft and Jetz 2007; Weigelt et al. 2016). Coevolutionary processes may also be important, whereas a larger number of species begets more diversity of parasites, predators, and symbionts, for example (Brown 2014). While silent on these processes, our simple theory still does a reasonable job in increasing our understanding of the fundamental processes involved. The "experimental toolbox" for biodiversity patterns that we have created is certainly still in its infancy, and we are excited to see what the addition of alternative mechanisms shaping biodiversity can tell us. We hope that the work reported here can serve as a starting point for others, to improve predictive capacity by expanding on the basic theory and its model implementations. We further suggest that the empirical and theoretical insights assembled here may have some applications in exploring possible effects of human drivers on the current state and future prospects for biodiversity, and its conservation at a global scale. These topics will be expanded on in chapter 6.

CHAPTER SIX

Conservation Applications

As stated in chapter 1, this book emerged out of a long-standing interest in synthesizing and explaining observed patterns of biodiversity (Stehli et al. 1969; Rohde 1992; Gaston 2000; Tittensor et al. 2010), but was equally motivated by concerns about the present status and future prospects for global biodiversity (Pimm et al. 1995; Worm et al. 2006; Tittensor et al. 2014; Beaugrand et al. 2015). There is little doubt that we are in the midst of a global biodiversity crisis, with local and global extinction rates ranging far outside the bounds of natural variation (Rockstrom et al. 2009; Dirzo et al. 2014; De Vos et al. 2015). How can our knowledge of global biodiversity patterns and our understanding of underlying processes and drivers help us to apprehend, project, and reverse the trajectory of large-scale biodiversity loss? That question forms a core motivation for our work, and will be explored in this chapter.

In reviewing this topic, we cannot hope to complete a synthesis of all existing work; such a comprehensive review undoubtedly would fill a volume on its own. The main thrust of our book concerns the development of an empirical and theoretical understanding of global biodiversity patterns and their underlying drivers, rather than designing approaches specifically targeted for conservation application. This is partly the case because our focus remains at regional to global scales, not at the finer spatial grain at which management interventions most often occur. Yet, there is a long history of biodiversity theory informing more applied work—for example, with respect to the Theory of Island Biogeography and the Unified Neutral Theory of Biodiversity and Biogeography (Mangel 2002). Here, we build on these earlier efforts and explore possible conservation applications of our empirical and theoretical synthesis.

Broadly, we see three major applications in conservation biology. The first concerns the present state of biodiversity on land and in the sea, and the development of a globally integrated empirical basis for conservation planning. Such an approach might use all available data from the four environmental realms identified in chapter 2, and work toward harmonizing conservation strategies across these disparate realms. The second major application concerns ongoing biodiversity change, and how a better understanding of fundamental drivers of diversity can help us prioritize mitigation of certain deleterious impacts over others. The third area of application concerns the future of biodiversity on this planet, and how an improved understanding of patterns, drivers, and changes in global biodiversity can help to project possible scenarios of long-term change, under a given rate of environmental change. We will briefly highlight these three potential applications in the following sections.

6.1. GLOBAL BIODIVERSITY HOTSPOTS
AND CONSERVATION PRIORITIES

One of the contributions we attempted to make in this book was to synthesize known patterns of global biodiversity across land and sea (chapter 2). Here, we examine what this empirical synthesis implies in terms of large-scale conservation prioritization on an increasingly fragile planet. An important generalization that emerged in both our current and previous work is that global patterns of species richness tend to be spatially correlated among taxonomically distant groups inhabiting the coastal, pelagic, and terrestrial (including freshwater) realms (Grenyer et al. 2006; Tittensor et al. 2010; Tisseuil et al. 2013). This generalization might also apply to the deep sea, but this cannot yet be ascertained, as only one group has been globally sampled and mapped (Woolley et al. 2016). This means that human impacts on these different habitats may produce correlated responses across species groups, at least when those impacts occur over the large scales examined here, and assuming that these correlations also hold for undersampled species groups not represented in our global overview.

This correlation structure across taxa is important for systematic conservation planning at the global scale and has been applied in a "hotspot" approach to conservation prioritization (Myers et al. 2000; Roberts et al. 2002; Worm et al. 2003; Selig et al. 2014). Within such a framework, areas of high species richness, the so-called biodiversity hotspots, are considered immediate conservation priorities, in order to safeguard the maximum number of species per unit of area that is protected. This prioritization scheme is often used in combination with other spatial data layers on biodiversity threats, endemism, rarity, or number of endangered species to provide a more comprehensive picture of actual conservation needs (Myers et al. 2000; Roberts et al. 2002; Orme et al. 2005; Selig et al. 2014). Of course, prioritizing total species richness as the metric of biodiversity might lead to different outcomes compared to alternative metrics such as functional diversity or phylogenetic distinctness (for example, Stuart-Smith et al. 2013). Regardless of the metric used, this prioritization scheme can be applied for high-level global decision making (Myers et al. 2000; Roberts et al. 2002; Selig et al. 2014), or be adapted for use at finer scales to inform local to regional conservation planning (for example, Cowling et al. 2003), while accounting for ongoing changes in human impacts and their dynamic effects on local biodiversity patterns (Pressey et al. 2007). Remarkably, this approach has never been applied to the planet as a whole, but only to terrestrial (for example, Myers et al. 2000; Orme et al. 2005) or marine (for example, Roberts et al. 2002; Selig et al. 2014) environments in isolation.

Here, we examine global richness patterns and biodiversity hotspots on land and in the sea together. We look at these patterns through two different lenses—namely, (1) total species richness, and (2) relative richness across taxa. For total richness, and combining all available data from chapter 2, fig. 6.1 shows the synthetic global pattern of total species richness summed across all well-sampled groups on land and in the sea. We observe that the land and coastal marine areas sustain higher richness than the pelagic ocean or the deep sea (fig. 6.1A), likely reflecting fundamental differences in temperature (surface being warmer than deep waters), habitat complexity and productivity (higher on land and in the coastal ocean than in pelagic and deep-sea habitats), and evolutionary history (more rapid diversification on land). On the other hand, some aspects of this strong pattern in absolute richness might reflect the biases of sampling well-known taxa and the effects of relative sampling effort, accessibility, and length of study on land and in coastal regions. Moreover, some features are likely shaped to an extent by species-rich groups such as land plants and coastal fishes (see chapter 2 for similarities and contrasts between individual taxa). Yet the observed land-sea gradient broadly matches estimated differences in total richness across these two realms (Mora et al. 2011), though there remains considerable uncertainty about total deep-sea biodiversity (Rex and Etter 2010).

The most species-richness cells in our global sample are consistently found in tropical coastal areas, where the highest observed richness of terrestrial and marine groups intersects. This is seen most prominently in Southeast Asia and in western South America, where global biodiversity hotspots converge between land and sea. Here, a conservation approach that integrates terrestrial and marine protected areas and other conservation measures may be particularly fruitful.

Conservation of species-rich hotspots becomes even more important, when considering that global patterns in sampled biodiversity also mirror similar gradients in human population density, environmental footprint, and associated impacts (fig. 6.1B). Such impacts do not concern the land or the sea in isolation, and concentrate in coastal areas, where more than half of humanity has chosen to settle. This human focus on the coast may at least in part be motivated by the rich portfolio of natural resources found there historically (Lotze et al. 2006). Due to this particular settlement pattern, human impacts thus tend to decline with increasing

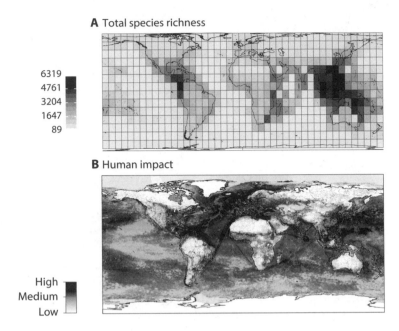

FIGURE 6.1. Total species richness and human impact. Shown is (A) the sum of species counts per cell for all taxa in chapter 2. (B) Human impacts expressed as cumulative relative impact. Data were derived for the land from the Human Footprint Map, accessed from NASA's Earth Observing System Data and Information System (EOSDIS) Socioeconomic Data and Applications Center (http://sedac.ciesin.columbia.edu/data/set/wildareas-v2-human-footprint-geographic/maps); marine aggregate human impact data were compiled from Halpern et al. (2015).

distance from the coasts, and both toward landlocked areas and toward the open ocean and deep sea (Vitousek et al. 1997; Imhoff et al. 2004; Halpern et al. 2008). This means that human impacts appear to be intensified in the world's most species-rich habitats, land and sea, imposing large risks on biodiversity there. The overlap between species richness and human impacts is particularly strong in East Asia, which is a major global biodiversity hotspot both on land as well as in the sea (fig. 6.1A), and home to 60% of the human population, with a massive footprint on local resource use (fig. 6.1B). Other marked hotspots of overlap in species richness and human impact, both land and sea, include Mesoamerica and the Caribbean, as well as East Africa including Madagascar (fig. 6.1). The available data strongly suggest that these areas should all be treated as marine-terrestrial conservation priorities at a global scale.

Another way of looking at global biodiversity is to describe hotspots of relative richness that are consistent across taxa, no matter what their absolute richness is. By normalizing species richness for each taxon and then averaging across all taxa present in each cell, we derive synthetic patterns of relative richness in which each group contributes equally to the total estimate, and the strong influence of particularly species-richness taxa is neutralized (fig. 6.2). Hotspot areas in fig. 6.2 are arbitrarily delineated by the top 10% of cells, in terms of average normalized richness across groups. On land, such consistent hotspots across sampled taxa emerged in the tropical Andes, central East Africa, and Southeast Asia (fig. 6.2A), all of which sustain tropical wet forests. We note that these cross-taxa hotspots broadly match the originally described hotspots for vascular plants (Myers et al. 2000), pointing again at the strong correlation between plant and vertebrate taxa. In the ocean, highest coastal diversity was observed in hotspots around the Indonesian-Philippines-Australian Archipelago (the "coral triangle"), southern India and Sri Lanka, southeast Africa and Madagascar, and the Caribbean (fig. 6.2B), all of which feature warm waters, extensive coastlines, and large islands that are located close to continental coastlines. In contrast, smaller island chains in the central Pacific (Micronesia and Polynesia) do not support the same number of species (fig. 6.2B). Pelagic species (fig. 6.2C) and deep-sea species (fig. 6.2D) provided a striking contrast to terrestrial and coastal marine species by showing much more uniform patterns of diversity across ocean basins, with hotspots at ~20 to 30 degrees latitude North or South in all oceans. Within these latitudes, the most diverse hotspots of pelagic species were typically located closer to the continents and along boundary currents such as the Gulf Stream and Kuroshio Current. This probably relates again to the availability of favorable habitat features of oceanic fronts often forming alongside boundary currents and along shelf breaks (Olson et al. 1994; Etnoyer et al. 2004; Worm et al. 2005). In the deep sea, hotspots of diversity for ophiuroids, the single taxon with global coverage, were seen in the North Atlantic, off Japan and New Zealand, and below coastal upwelling zones in southwest Africa, South America, and the California Current (fig. 6.2D). The North Atlantic hotspot was also seen in other taxa that were more regionally sampled, such as mollusks (Tittensor et al. 2011). When comparing relative richness patterns across all sampled realms and taxa, the Western Pacific Rim emerged as particularly prominent for terrestrial, coastal, and pelagic taxa, including hotspots around Japan, the Philippines, and Australia (fig. 6.2). The North Atlantic, particularly off Europe and North America, showed regional hotspots for both pelagic and deep-sea taxa; similar overlap was seen in the southern hemisphere off South Africa and New Zealand. Land and deep-sea species richness hotspots coincided around eastern South America.

This highly nonrandom distribution of species richness within and across taxa and realms is an important feature to inform systematic conservation planning on regional to global scales (Margules and Pressey 2000; Myers et al. 2000). Specific habitat features tend to harbor the highest observed diversity across species groups; these include tropical moist forests on land, tropical

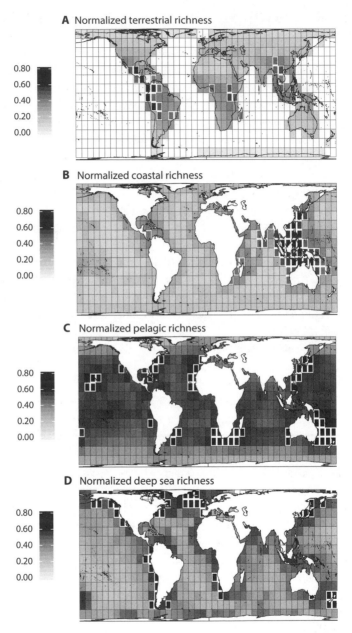

FIGURE 6.2. Hotspots of species richness across taxa. Shown are normalized patterns of species richness across all species groups in each of the four major realms: (A) land (4 taxa), (B) coastal (9 taxa), (C) pelagic (7 taxa), and (D) deep sea (1 taxon). The 10% most species rich cells are outlined in bold white. Richness data were derived from sources in table 2.1.

archipelagos in the coastal oceans, frontal areas at intermediate latitudes for pelagic species, and areas of high export production for deep-sea species. Clearly, areas and regions that harbor these critical habitat features, and that show overlap between different taxa, should receive priority attention under a hotspot approach, due to their importance across species and their unique global significance for biodiversity.

While the hotspot approach is an intuitive first-order method for spatial conservation, it clearly cannot be sufficient on its own, as large parts of the planet will not be represented by it, such as land and coastal areas at higher latitudes (Kareiva and Marvier 2003). Likewise, when prioritizing endemic or threatened species (Orme et al. 2005), or looking at functional or phylogenetic diversity (Safi et al. 2011; Devictor et al. 2010; Stuart-Smith et al. 2013), other hotspot patterns may emerge and need to be taken into account to comprehensively safeguard biodiversity at a global scale. Furthermore, real-world conservation efforts also reflect the changing priorities that society, or conservation organizations, has at a given time, focusing on particular species groups or habitats of value to people. Merging such dynamic socioeconomic preferences with information on the biological patterns documented here remains an interdisciplinary frontier in conservation planning. Finally, another aspect that is becoming increasingly prominent is rapid environmental change and the effects this may have on species distribution, richness, and conservation outcomes (Pressey et al. 2007). This is briefly described in the following section.

6.2. OBSERVED BIODIVERSITY CHANGE AND ITS DRIVERS

As discussed in chapters 2 and 3, biodiversity has been a dynamic entity over the past 4 billion years. Yet, at the present time, biodiversity is changing much faster than throughout most of Earth's history, driven by unprecedented environmental change brought about by manifold human impacts. In this section, we briefly review what is known about present biodiversity change, and how it relates to our previously discussed understanding of fundamental drivers. We suggest that an improved understanding of biodiversity drivers gained in chapter 3 can help us to prioritize the mitigation of certain human impacts over others, and to project future scenarios of biodiversity change from process-based models (see also the following sections).

As biodiversity continues to change and evolve in response to environmental change, it is paramount for conservation biologists to gain a better understanding of the direction of change, and the main drivers that effect this (Sala et al.

TABLE 6.1. Average Direction and Model Implementation
of Global Impacts on Biodiversity

Driver of diversity	Direction of human impact				Model implementation	Source
	Land	Coast	Pelagic	Deep sea		
Temperature	↑	↑	↑	↑	Temperature	Burrows et al. 2011; Allen et al. 2014
Habitat area	↓	↓	→	→	Community size	Bender et al. 1998; Worm and Lenihan 2013
Productivity	↑	↑	↓	↓	Community size	Nemani et al. 2003; Boyce and Worm 2015
Disturbance	↑	↑	↑	↑	Community turnover	Watling and Norse 1998; Dirzo et al. 2014
Connectivity	↓	↓	→	→	Dispersal rate	Fahrig 2003; Cowen et al. 2006

Note: Shown are major drivers of biodiversity that are affected by various human impacts; the approximate direction of global impacts is indicated by arrows (↑ increasing, ↓ decreasing, → minor change). Model implementations (see chapter 4) of key parameters that can be adjusted to simulate change in human impacts are listed. References highlight documented global impacts and their direction.

2000; Thomas et al. 2004; Tittensor et al. 2014). In our empirical analyses and theoretical models, temperature (affecting species distributions, community turnover rates, and speciation rates) as well as habitat area and productivity (driving changes in local community size) emerge as strong predictors of species richness through space and time. This is significant, because human activities increasingly affect and drive these variables (Waters et al. 2016). There is now ample evidence that human-induced changes in temperature, habitat area and complexity, productivity and disturbance, as well as changes in connectivity among ecosystems, are affecting biodiversity worldwide (table 6.1). Arguably, these are among the most severe human impacts, precisely because they affect fundamental drivers of biodiversity. These drivers have been much studied, and their effects have been modeled using various approaches, In this section, we will very briefly explore such processes as they unfold across various ecosystems, and assess to what degree observed changes in species distribution and diversity may relate to predictions made by our theory.

6.2.1. Effects of Temperature and Climate Change

We are living through an era of rapid global climate change (Allen et al. 2014), which is causing both regional- and planetary-scale changes in weather patterns, temperature, and ice cover (Barnett et al. 2001; Mora et al. 2013) and further affects sea level, thermal stratification regimes, atmospheric and oceanic circulation patterns, ocean pH, oxygen content, and productivity (Richardson and Schoeman 2004; Sarmiento et al. 2004; Bryden et al. 2005; Behrenfeld et al. 2006; Polovina et al. 2008; Boyce et al. 2010; Lewandowska et al. 2014; Boyce and Worm 2015; Waters et al. 2016). As such, global climate change is predicted to have complex effects on species distributions and diversity (Sala et al. 2000; Brander 2010; Sydeman et al. 2015).

Clearly, temperature change in particular directly relates to species richness empirically (chapter 2) and in our theory and model (chapters 4 and 5). Sustained changes in temperature under global warming will affect species' thermal niche distributions (ecological timescale), individual metabolic rates and community turnover (ecological timescale), and speciation and diversification (evolutionary timescales). Here, we briefly review what is known empirically about the effects of temperature changes on observed diversity patterns.

In the oceans, long-term fish and plankton monitoring programs have provided particularly compelling evidence for temperature- and climate-driven changes in species diversity (Hays et al. 2005; Perry et al. 2005; Pinsky et al. 2013). Observations most consistently indicated an increase in species richness in temperate regions, as warm-adapted species invade from the subtropics and more than compensate for cold-adapted species that are displaced to higher latitudes. The net effect of such temperature-induced compositional changes on species richness can be surprisingly large: for fish, an almost 50% increase in the number of species recorded per year in North Sea bottom trawl surveys was documented between 1985 and 2006 (Hiddink and ter Hofstede 2008). This change correlated tightly with increasing water temperatures. Similar trends have been found in the United Kingdom's Bristol Channel, where fish species richness increased by 39% from 1982 to 1998 (Henderson 2007). In both cases, increases in richness were mainly driven by invasion of small-bodied southern species. Similar changes have been observed for terrestrial taxa, where long-term monitoring has historically been somewhat more straightforward. For example, the species richness of British butterflies (Menéndez et al. 2006) and epiphytic lichen in the Netherlands (van Herk et al. 2002) have increased with warming over time.

Some polar regions have seen a similar pattern of slow invasion by temperate species. However, only few observations on net changes in species richness are available (Wassmann et al. 2011)—for example, surveys of Arctic macrobenthos suggest slow increases in species numbers at sites that are accessible to larval advection from southern locations (Węsławski et al. 2011). Large uncertainties remain due in part to low sampling effort (Wassmann et al. 2011), and in part due to the complexity of this highly seasonal environment, and the compounding effect of changes in sea ice, salinity, stratification, runoff, and acidity (Michel et al. 2012).

Under rapid warming, tropical regions become increasingly heat-stressed and may decline in richness, particularly where there is a loss of foundation species such as trees or corals. For example, coral reefs have experienced mass bleaching where sea temperatures have exceeded long-term summer averages by more than 1°C for several weeks (Hoegh-Guldberg 1999; Donner et al. 2005; Hughes et al. 2017). The loss of sensitive coral species causes secondary changes of reef-associated or obligate fauna and flora (McClanahan et al. 2001). For reef fish specifically, studies indicate large changes in species composition after bleaching events, and a decline in species diversity that is linearly related to disturbance intensity (Wilson et al. 2006), but see Bellwood et al. (2006) for an exception. Other habitat-forming species, such as seagrasses, mangroves, and some seaweeds, also face elevated extinction risk due to warming and sea-level rise (Polidoro et al. 2010; Short et al. 2011; Harley et al. 2012), with consequences for communities dependent on these habitats.

On land, the effects of warming are also complicated by changing patterns of rainfall, which could enhance or reduce diversity (Bellard et al. 2012). Empirical evidence for lizards (Sinervo et al. 2010) and amphibians (Pounds et al. 2006) suggest, however, that climate-related extinctions are already under way on land. More generally, warming will increase metabolic rates in these and other ectotherm species, and this increase will be disproportionally large in tropical species due to the nonlinear relationship between temperature and metabolic rate (Dillon et al. 2010). Such increases likely have substantial physiological and ecological impacts including an increased need for food, increased rates of evaporative water loss in dry environments, behavioral change, changes in tropical soil respiration, and altered food web dynamics (Dillon et al. 2010). In the long term, as discussed at length in chapter 3, metabolic theory also predicts increased rates of molecular evolution and speciation (Allen and Gillooly 2006), which could ultimately foster biological innovation and novel adaptations.

In summary, as the planet rapidly warms, and as a first approximation, the tropics may lose diversity, temperate regions show species turnover and increases in net diversity, whereas polar environments so far mostly show declines in ice-dependent species and some invasion of subpolar taxa with unclear effects on net diversity. Constraints to range shifts and expansions appear to be less important in the oceans than on the land. In the North Sea, for example, the average rate of northward range shift was 2.2 km a^{-1} (Perry et al. 2005), more than three times faster than observed range shifts in terrestrial environments, which reportedly average 0.6 km a^{-1} (Parmesan and Yohe 2003). Likewise, a meta-analysis of species range shifts showed that marine species fill their thermal niches more fully and move more readily at both cold and warm range boundaries compared with than terrestrial species (Sunday et al. 2012). These findings may not be surprising, given the absence of hard physical boundaries in marine, and particularly pelagic, environments. There are, however other potential constraints, such as light availability for corals at higher latitudes, or lower oxygen for fish in warmer waters, that might prove to set unexpected boundaries to dispersal (Kleypas 2015). Similarly, changes in habitat area or productivity could constrain realized changes in species richness relative to anticipated outcomes based solely on thermal change, as discussed later.

6.2.2. Changes in Habitat Area and Productivity

In addition to its effect on temperature, global environmental change has other important consequences, such as changes in habitat quality, quantity, and productivity, that directly affect the carrying capacity of ecological communities (Vitousek 1994; Vitousek et al. 1997).

On land and in coastal environments, the physical alteration of habitats often reduces the effective area available for colonization—for example, where coastlines are modified for human infrastructure (Airoldi and Beck 2007) or forests are cleared for agriculture (Ellis et al. 2010; Newbold et al. 2015). Specifically, between the years 1700 and 2000 the terrestrial biosphere made a transition from mostly wild to mostly human-dominated habitats, passing the 50% mark early in the twentieth century (Ellis et al. 2010). Interestingly, some of the associated losses in native species richness have been compensated for by species invasions, such that overall observed richness has not changed dramatically in some cases, though composition has (Ellis et al. 2012). The rate of habitat destruction, however, will likely accelerate in the future, as a function of human population growth and increasing affluence (Godfray et al. 2010), with a significant projected "extinction debt" for terrestrial species (Tilman et al. 1994). Similar historical trends have been documented in marine coastal environments (Pandolfi et al. 2003; Lotze et al. 2006), with increasing industrialization and habitat conversion moving outward from heavily populated coastlines toward the open ocean (McCauley et al. 2015). In addition, habitat fragmentation is occurring alongside habitat changes and has far-reaching effects on biodiversity (Fahrig 2003), with the potential for synergistic impacts from these two habitat-related stressors (Bartlett et al. 2016).

Just as the effects of habitat alteration, changes in productivity can be equally important, as they affect community size and composition, and consequently diversity. For example, a temperate reef fish community in southern California initially saw increased species richness with gradual warming over time (Holbrook et al. 1997). Yet, sudden warming events also led to a decline in productivity, 80% loss of large zooplankton biomass, and recruitment failure of many reef fish. This may explain why total biomass declined significantly, and total species richness also declined by 15 to 25% at the two study sites despite increases in temperature (Holbrook et al. 1997). This example illustrates that predictions of changes in species composition and diversity that are based solely on temperature can be misleading, if concomitant changes in habitat area, productivity or other drivers are ignored. Such countervailing influences are particularly important in the oceans, as warming waters are becoming more stratified, reducing the delivery of nutrients from deeper waters (Behrenfeld et al. 2006), which has already depressed plankton biomass, particularly in pelagic ecosystems far from shore (Polovina et al. 2008; Boyce et al. 2014). A recently observed signal of changes in phytoplankton biomass on fish production at the early life stage suggests that these ongoing changes could have far-reaching effects on species communities at multiple trophic levels (Britten et al. 2016). On land and in the coastal ocean, however, productivity may be broadly increasing due to anthropogenic mobilization of nitrogen and phosphorus, which stimulates primary production, with variable effects on species richness (Worm et al. 2002; Stevens et al. 2004). The increasing magnitude and global scale of atmospheric nitrogen deposition in particular is bound to have far-reaching effects on the productivity and diversity of land and ocean communities worldwide (Jefferies and Maron 1997; Sala et al. 2000; Duce et al. 2008). In the terrestrial ecosystem, however, there is added uncertainty, associated with other processes, such as carbon cycle feedbacks due to elevated $CO2$, as well as changes in growing season length, and land use (Erb et al. 2016). Despite these complexities, the overall signal emerging from historical time series is one of increased primary production in terrestrial (Nemani et al. 2003; Campbell et al. 2017) and reduced production in marine environments (Gregg et al. 2003; Boyce and Worm 2016), with so far unknown effects on global biodiversity patterns on land and in the sea.

6.2.3. Disturbance and Exploitation

In addition to the effects of temperature, habitat, and productivity change, direct disturbance and removal of species by logging, hunting or fishing can also represent a dominant impact on species richness. For example, the widespread "wild meat" hunt on land (Milner-Gulland et al. 2003), or the estimated annual fish catch of ~100 Mt from the oceans (Pauly and Zeller 2016) can affect the regional patterns of species richness (Worm et al. 2005; Effiom et al. 2013). Indeed, observed richness of large fish (>25 cm length) on standardized underwater visual surveys is depressed in most fished areas, but elevated in well-enforced marine protected areas worldwide (Edgar et al. 2014). A model of species richness for all reef-associated fish, large and small, showed the typical pattern of maximum richness in the western tropical Pacific (Edgar et al. 2014). In contrast, observed richness for large fish showed a subtly different pattern, peaking at remote sites in the central Pacific, around French Polynesia and the Line Islands (Edgar et al. 2014). Similarly, an analysis of heavily exploited tuna and billfish species sampled by a globally operating long-line fleet indicated large-scale changes in species richness patterns from the 1950s to 1990s, and overall declines in average richness in the Atlantic and Indian Ocean, but not the Pacific (Worm et al. 2005; Worm et al. 2010). Changing richness patterns were at least in part due to declining range extent in a number of species, such as three species of bluefin tuna (Worm and Tittensor 2011). Likewise, a historical data set of coastal biodiversity showed that 7% of the species for which data were available went regionally extinct over the last 1000 years (Lotze et al. 2006; Worm et al. 2006). Clearly, fishing is reshaping patterns of local and regional richness for exploited groups, as well as the spatial scaling of biodiversity (Tittensor et al. 2007), and this change might skew our perception of what the baseline patterns of biodiversity are (Pauly 1995; Baum and Myers 2004).

Despite these large regional changes, global extinctions remain relatively few in the oceans. Only two marine fish species are thought to have gone globally extinct over recent human history, although undersampling in reef and deep-water environments may hide unrecognized extinction events (Roberts 2002; Reynolds et al. 2005). However, four marine mammals and four mollusks are known to be lost, as well as a number of freshwater fish and one anadromous species, the New Zealand grayling, according to the International Union for Conservation of Nature (IUCN) Red List (Worm and Lenihan 2013). This means that the overall global richness of marine life is likely still similar to preindustrial times, but that local and regional patterns of community structure, biomass, and diversity have changed, sometimes dramatically, especially for heavily exploited species (Lotze et al. 2006; Estes et al. 2011). This is reflected in the IUCN assessments, which listed between 9% (for bony fish) and 38% (for marine mammals) of non-data-deficient species as threatened by extinction in 2013 (Worm and Lenihan 2013). On land, in contrast, more than 700 extinctions have already occurred, many of them due to exploitation or human-mediated invasion by exotic species, with tens of thousands threatened by extinction in the near future (Barnosky et al. 2011). While this situation does not yet reach the magnitude of previous mass extinction events, human impacts are undoubtedly reshaping current and future patterns of biodiversity (Barnosky et al. 2011; Waters et al. 2016), with documented effects on trophic structure and ecosystem functioning (Worm et al. 2006; Estes et al. 2011).

6.2.4. Model Implementation

In summary, many (but not all) of the human impacts that are affecting biodiversity on land and in the sea relate directly to key processes that are driving observed patterns of species richness—namely, temperature, habitat area and connectivity, productivity, and disturbances. As such, it seems prudent to further investigate these impacts as we attempt to mitigate biodiversity change and loss. Where empirical observations are scarce or unavailable—for example, in poorly sampled regions, or with respect to the future—models can help us to explore possible scenarios of biodiversity change under a given impact scenario. While there are many other process-based models that can potentially shed light on human impacts in biodiversity, we suggest that the theoretical framework presented in this volume may, with some modification, be useful to further explore and project the likely long-term consequences of these impacts, and their effects on species richness now and in the future (see table 6.1). As discussed in chapters 4 and 5, changes in surface temperature are readily incorporated into the model via a global temperature grid that can be adjusted to reflect observed or projected changes.

Changes in suitable habitat area correspond directly to changes in the community size parameter J, and could thus be potentially assessed alongside the effects of temperature. Increasing habitat fragmentation limits connectivity among neighboring communities, and could be assessed through modification in dispersal rate m, adjusting the connectivity between local communities in our metacommunity model (see table 6.1). Changes in disturbance or exploitation rate can be implemented as changes in local community turnover. As such, our metacommunity model may provide an experimental toolbox for projecting the complex effects of single or combined environmental impacts and changes. As an example, we will illustrate in the following the use of our model framework for exploring scenarios of future climate change, in order to understand possible long-term effects on biodiversity.

6.3. PROJECTING BIODIVERSITY CHANGE FROM THEORY

Unprecedented rates of environmental changes are making biodiversity conservation a moving target, complicating present and future conservation efforts (Pressey et al. 2007; Worm and Lotze 2015). As such, conservation planning increasingly uses empirical data synthesis and process-based models to project possible changes in biodiversity, and the effects it may have on conservation outcomes. As spatial data on recent changes in species richness patterns are typically not available at a global scale (but see Worm et al. 2010; Dornelas et al. 2014; Newbold et al. 2016 for exceptions), species distribution models have been utilized most commonly to fill this gap. These models typically reconstruct biodiversity patterns by overlaying species ranges derived from habitat requirements, such as thermal tolerances, for example. Such models can be used to forecast future changes in species richness, given projected scenarios of climate and habitat change (Cheung et al. 2009; Kaschner et al. 2011). Here, we briefly review existing modeling approaches and then present a new, and possibly complementary, approach based on our global metacommunity model presented in chapters 4 and 5.

6.3.1. Projections from Species Distribution Models

Species distribution models (SDMs) are currently the most widely used modeling tool to project the possible future impacts of environmental change on biodiversity (Franklin et al. 2013). Terrestrially, there exists a voluminous literature on projecting biodiversity changes using SDMs (see, for example, Urban 2015); we don't attempt to survey it further here. The vast majority of projections for future biodiversity scenarios have been derived from SDMs assuming static thermal tolerances and thermal niches. For example, Molinos et al. (2016) recently used observed temperature tolerances and habitat preferences for 12,796 coastal and pelagic species represented in Aquamaps, an online resource that maps out predicted species ranges in the marine environment (www.aquamaps.org). The authors projected overall increases in average biodiversity under both moderate and high greenhouse gas emissions, as implemented in the representative concentration pathways RCP 4.5 and RCP 8.5 (fig. 6.3). Increases in species richness were projected at high latitudes, matching empirical observations (see section 6.2.1, earlier). But these regional increases were partly offset by declining richness in tropical regions, especially at the higher emission scenario (fig. 6.3); this also led to a partial flattening of latitudinal richness gradients (Molinos et al. 2016). It is important to remember that these changes were entirely based on assumed contraction or expansion of predicted species ranges; neither speciation nor extinction processes were included, assuming no evolutionary change. Tropical losses of biodiversity, for example, were driven by the displacement of tropical species to higher latitudes, under assumed constant temperature tolerances, and no evolution. Parallel work for fishes (Cheung et al. 2009) and marine mammals (Kaschner et al. 2011) produced similar results for individual species groups,

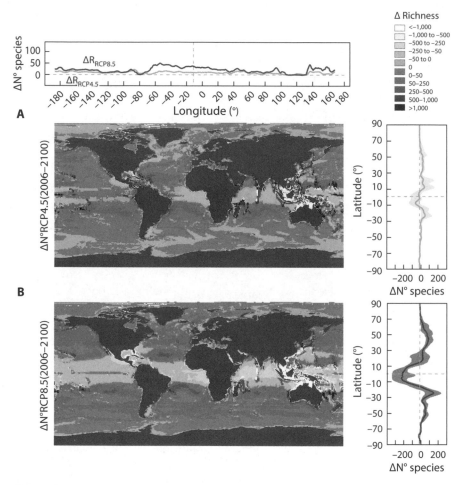

FIGURE 6.3. Future projections for marine species richness from a species distribution model. Shown are projected changes in absolute species richness ($n = 12,796$ species) from 2006 to 2100 given two representative concentration pathways (RCPs) devised by the Intergovernmental Panel on Climate Change (IPCC), (A) RCP 4.5 and (B) RCP 8.5. Black contour lines correspond to countries' exclusive economic zones (EEZs). Changes to average latitudinal and longitudinal gradients (solid line) with their 25% and 75% quartiles (shaded areas) are also shown. Note that the high-emission scenario results in much greater species losses, particularly in the tropics. After data in Molinos et al. (2016).

with some more recent work including more realistic dispersal constraints into the SDMs (Cheung et al. 2015).

Beaugrand and coworkers used a more theoretical approach of modeled distributions for randomly generated pelagic "pseudospecies," mimicking present-day distributions of pelagic zooplankton groups (Beaugrand et al. 2015). Their results showed patterns similar to the previously mentioned SDMs. The authors quantified projected changes relative to those that might have occurred between the last glacial maximum and the present. For example, under severe global warming (RCP 6.0 and RCP 8.5), between 50 and 70% of the global ocean was projected to show a magnitude of biodiversity change not seen since the last glacial maximum (Beaugrand et al. 2015).

Other approaches that sought to model future biodiversity have been based on observed species-environment relationships from electronic tracking data (Hazen et al. 2013), or observed species richness-environment relationships from observational data (Whitehead et al. 2008). What connects all of these contemporary approaches is that they do not include evolutionary dynamics. Species distribution models may capture the short-term rearrangement of species that are thought to move along with their temperature niche, but do not capture the effects of changes in temperature on the processes that generate biodiversity, nor the evolutionary plasticity that allowed species to survive previous environmental perturbations. Microevolutionary patterns of adaptation to changes in the environment are becoming increasingly obvious in empirical studies (Hoffmann and Sgrò 2011). For example, it has been shown experimentally that stickleback evolve quickly from marine forms to freshwater forms, with one major aspect being rapid adaptation to low temperatures (Barrett et al. 2010). Similar patterns of rapid evolutionary adaptation have recently emerged in marine phytoplankton (Schlüter et al. 2014) and corals (Palumbi et al. 2014). This recent work collectively suggests that changes in ocean temperature and acidity provide major evolutionary challenges along which populations diversify within species, and most likely across species and lineages. Adaptation to very high temperatures, however, appears particularly challenging for many taxa and could represent a hard evolutionary barrier. Araújo et al. (2013) analyzed thermal tolerances of a large number of terrestrial ectotherm ($n = 697$), endotherm ($n = 227$), and plant ($n = 1816$) species worldwide, and showed that tolerance to heat is largely conserved across lineages, while tolerance to cold is more malleable and varies more widely between and within species. It is interesting that the evolutionary processes that may have shaped global diversity patterns over millennia are also observable in real time (Barrett et al. 2010; Hoffmann and Sgrò 2011; Palumbi et al. 2014; Schlüter et al. 2014). In terms of the approach in this volume, our eco-evolutionary metacommunity model allows for such evolutionary processes to occur, while also tracking changes in habitat, disturbance regimes, dispersal, and thermal adaptation.

6.3.2. Projections from a Neutral-Metabolic Model

In order to experiment with the utility of our approach for understanding dynamic changes owing to human impacts, we utilized the richness patterns predicted by our

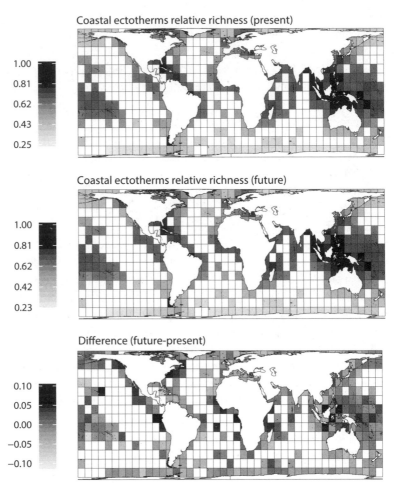

FIGURE 6.4. Future projections for coastal marine species richness from a neutral-metabolic metacommunity model. Shown are relative changes in species richness from 2006–2010 to 2091–2100 given projected climate change effects on surface temperature and precipitations forced by the representative concentration pathway RCP 8.5.

global metacommunity model (see chapter 5) to explore possible future changes in species richness from projected global warming and altered moisture regimes for coastal ectotherms (fig. 6.4) and land plants (fig. 6.5). This is different from the previously described modeling approaches in that no species identities are tracked, and no ecological differences are assumed; we recognize the limitations here but note that this allows us to explore processes that are typically left aside. Projected changes in this framework are due to thermal effects on speciation and turnover

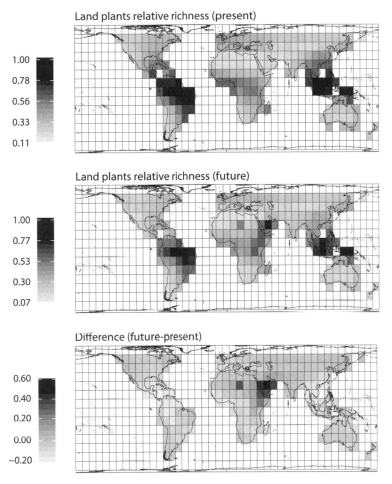

FIGURE 6.5. Future projections for land plant species richness from a neutral-metabolic model. Shown are relative changes in species richness from 2006–2010 to 2091–2100 given projected climate change effects on surface temperature and precipitations forced by the representative concentration pathway RCP 8.5.

rates, as well as effects of changes in moisture regime on productivity and hence community size on land. These are mechanisms that are typically not included in other modeling approaches, which tend to focus on thermal niches. Standard settings for the base parameters ($m = 0.1$, $v = 0.01$) were used as in previous simulations. For environmental drivers, we used RCP 8.5 from the Institute Pierre Simon Laplace (IPSL) Coupled Model Intercomparison Project (CMIP5) bias-corrected model runs (http://icmc.ipsl.fr/index.php/icmc-models/icmc-ipsl-cm5).

We extracted mean modeled land and sea surface temperature in 2091–2100 and proportional changes in precipitation between 2006–2010 and 2091–2100. We did not use predicted precipitation change directly in our model, but rather translated this information into changes in wet days, assuming linear scaling between annual precipitation and number of wet days, which have are a good global predictor of species richness for terrestrial plants (Kreft and Jetz 2007).

One critical aspect of this exploration is that the effects of temperature on community turnover and speciation are calculated relative to other local communities rather than pinned to an absolute value (that is, the summed probabilities for all individuals in all communities acting within each time step is equal to one). This means that while the model, in its present form, can be used to examine the relative change in spatial patterns of species richness, it cannot be used to examine absolute changes in richness values (though with further modification this may be feasible). Note also that our metacommunity model runs to a dynamic equilibrium state. Hence, the question we are asking is: How might the long-term ecological and evolutionary changes effected by changes in temperature and moisture alter global richness patterns and gradients relative to the present day?

We ran future explorations for coastal ectotherms and land plants, contrasting marine and terrestrial patterns of response to projected climate change. Changes in the future richness of ectothermic coastal species (see fig. 6.4) were calculated in response to projected changes in temperature only, as habitat effects via changes in coastline length could not be parameterized. As in the species distribution models discussed earlier, normalized species richness (relative to the observed global maximum richness) increased in most cells, particularly those projected to see large increases in temperature, such as those in the northwest Atlantic (see fig. 6.4). Some decreases in relative richness were observed along European and Caribbean coastlines, though.

On land, vascular plant richness according to empirical analyses (Kreft and Jetz 2007) and our theoretical models (chapter 5) is affected by changes in both temperature and moisture regimes. Both of these variables are tracked by Earth System Models and can hence be included in our projections (see fig. 6.5). Our model projected reduced relative richness in the two present hotspots, the tropical Andes and South America, but also in Europe and West Africa. Large increases in regional richness were projected for the Arabian Peninsula, for example, following projected increases in precipitation (see fig. 6.5). Some major features of these patterns of change were also reproduced by projections based simply on extrapolation of the statistical relationships between temperature, moisture, and plant species richness (Sommer et al. 2010).

These results should not be viewed as solid predictions but rather simple explorations based on changes in only one or two factors, and ignoring many other aspects biological and environmental change. Similar shortcomings, however apply to other modeling approaches that have been used for future projections; ultimately, we would search for commonalities and differences that emerge across a variety different modeling approaches.

We do see our approach as complementary to existing modeling tools, and an illustration of how evolutionary processes could be accounted for in global biodiversity scenarios. Our approach allows us to explore possible long-term ecological and evolutionary consequences of climate change, but may also potentially be applied to integrate the shorter-term effects of changes in community size as driven by habitat loss and changes in productivity (see table 6.1). As such, the theoretical models developed in this book may represent an interesting contrast to the commonly used SDMs that bypass possible evolutionary dynamics, or to the projections of biodiversity loss due to climate change that rely on statistical relationships such as the species-area curve (Thomas et al. 2004), which have been criticized by some (He and Hubbell 2011).

Future explorations could also examine transient (nonequilibrium) dynamics that may arise in response to projected environmental changes. Possibly, and with some more dedicated development, the ideas and models developed here might be expanded and in principle find application in environmental forecasting and scenario building.

6.4. THE FUTURE OF BIODIVERSITY

As discussed throughout this volume, biodiversity patterns are not a static feature. Both in recent decades, and certainly throughout Earth's history, the global magnitude and distribution of biodiversity has been dynamically changing in response to various environmental drivers, many of which are now affected by human activities. This means that the future of biodiversity is in a very real sense in our own hands, and future trajectories will largely depend on how we choose to constrain or manage the cumulative impacts that arise from our actions. This aspect of biodiversity change is not easily captured by ecological theory, as it is largely dependent on societal choices and interactions. Fortunately, there is growing awareness of this simple fact, and global commitments to halt the loss of biodiversity are being implemented as we write this.

When tracking changes in these commitments, however, it is often found that increases in pressures on biodiversity outweigh increasing societal responses to stem the loss of biodiversity (Butchart et al. 2004; Butchart et al. 2010; Tittensor et al. 2014). In a recent update on this topic, progress toward the 20 "Aichi" targets—a set of internationally agreed goals for biodiversity conservation under

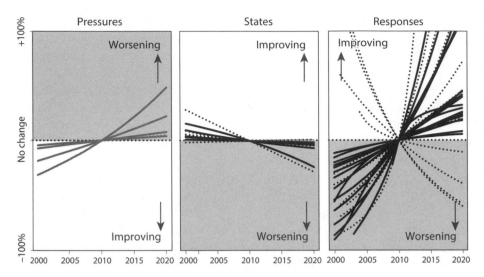

FIGURE 6.6. Global trends in biodiversity pressures, biodiversity states, and societal responses to biodiversity loss. Each line represents a unique response variable and its estimated significant (continuous) or nonsignificant (dotted) trend, extrapolated to 2020. Lines represent trends relative to 2010 value (horizontal dotted black line). Redrawn; after Tittensor et al. (2014).

the Convention on Biological Diversity—was tracked using 55 global indicators of biodiversity pressures, states, benefits, and responses to biodiversity loss (fig. 6.6). Results indicate that the rate of biodiversity loss may have slowed for some indicators, but that about half still show significant declines. The reason for this is undoubtedly that pressures are uniformly rising, and this is only partly compensated for by growing responses to biodiversity loss (Tittensor et al. 2014). Clearly, the biogeography of future biodiversity will hinge to a significant degree on changes in these societal parameters, emerging from our individual as well as collective choices. With the upcoming 2030 target for the United Nations Sustainable Development Goals, and beyond that the need to preserve biodiversity and ecosystems, the degree to which we can alter individual and societal choices that affect our shared natural environment represents an important research topic—perhaps one of the most important facing us.

CHAPTER SEVEN

Conclusions

A striking and primary feature of our planet is the bewildering diversity of species that inhabit it. As far as we know, the geographic distribution of these species has always been highly nonrandom (Stehli et al. 1969; Yasuhara et al. 2012; Mannion et al. 2014), and spatial variation in biodiversity along latitudinal, altitudinal, and moisture gradients has long been recognized by naturalists (Gaston 2000). Here, we approached these general patterns from an empirical and a theoretical perspective, in an attempt to gather a more comprehensive understanding of the processes that control the distribution of species richness at the global scale. Our approach was driven by two lines of inquiry—specifically, (1) synthesizing, testing, and contrasting observed patterns, hypothesized drivers, and environmental predictors across all available taxa on land and in the ocean on the same global-scale grid; and (2) integrating aspects of neutral, metabolic, and niche theories into a synthetic *neutral-metabolic-niche* (NMN) theory of biodiversity (fig. 7.1A) that generates a surprisingly accurate picture of global richness patterns from few underlying processes.

7.1. SUMMARY OF MAJOR FINDINGS

In our empirical synthesis, we found clear evidence that global biodiversity organizes into distinct patterns within four major biogeographic realms: coastal, pelagic, deep ocean, and land. Taxonomically distinct species groups tended to show similar patterns of biodiversity at large scales within each of these four realms (chapter 2). These patterns included steep latitudinal gradients in species richness from tropical latitudes (but not always the equator) to the poles for most coastal and terrestrial groups, and broader peaks of species richness across subtropical or temperate regions for pelagic and deep-sea taxa, respectively. Interestingly, we found that patterns appeared more robust and predictable at progressively higher levels of the taxonomic hierarchy (Class level and up), and were more likely to break down at the Order or Family level. When correlating species-richness patterns to environmental predictors, we found that ambient thermal

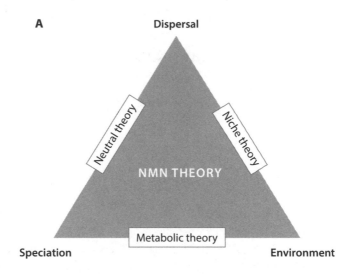

A

Dispersal

Neutral theory

Niche theory

NMN THEORY

Metabolic theory

Speciation   Environment

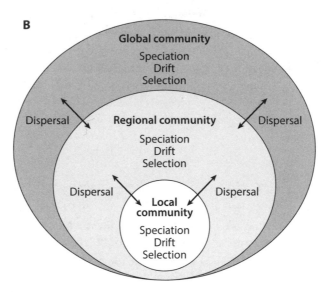

B

Global community
Speciation
Drift
Selection

Dispersal   **Regional community**   Dispersal
Speciation
Drift
Selection

Dispersal   **Local community**   Dispersal
Speciation
Drift
Selection

FIGURE 7.1. Synthesis in ecological theory. (A) The theory that is presented in this volume considers these fundamental processes by combining aspects of neutral, metabolic, and niche theory into a unified framework (NMN theory) that can be used for studying large-scale biodiversity patterns. While previous theories focused on dispersal and speciation (neutral theory), or dispersal-environment relations (niche theory), or how evolution is shaped by the environment (metabolic theory), we are modeling all three processes and their relationships in a synthetic, mathematically explicit framework. (B) In a previous conceptual synthesis, Vellend (2010, 2016) identifies four fundamental processes that shape local, regional, and global community structure and biodiversity: speciation, random drift, and selection shape local, regional, and global species pools, whereas dispersal drives the exchange of species across these pools.

energy (typically measured as average surface temperature) was the only factor that empirically correlated with areas of high diversity across all realms we investigated (chapter 3). Notable exceptions included cold-adapted endotherms such as pinnipeds, and deep-sea species, which occur in environments that are near-uniformly cold, with very limited temperature variation. Within each realm, we found that variables related to habitat area and productivity were also important (see table 3.6), both of which constrain the number of individuals in a community.

Based on this empirical synthesis, we attempted to devise a body of theory that might explain observed biodiversity patterns within and across taxa (chapter 4). The idea was to examine the ecological and evolutionary processes by which empirically documented predictors could affect diversity—for example, by changing community size or turnover, influencing the speed of evolution, or constraining dispersal. From this theoretical framework, we built a global meta-community model that could simulate these processes, singularly or in combination, and generate global biodiversity patterns, thus enabling us to investigate their relative influence. This work entailed a synthesis of three largely separate bodies of ecological theory—namely, neutral theory (Bell 2001; Hubbell 2001), which includes dispersal and speciation; metabolic theory (Allen et al. 2002; Brown et al. 2004), which parametrizes environmental constraints on metabolic activity and speciation rates; and niche theory (Hutchinson 1957; Pearman et al. 2008; Beaugrand et al. 2013), which links the environment to constraints on dispersal (fig. 7.1A). Hence, we developed a synthetic theory that attempted to reconcile previously idiosyncratic patterns of global biodiversity on land and in the ocean, and produced generally strong fits to the empirical data via our model implementation (chapter 5).

Somewhat surprisingly, this theory suggests that only two variables are required to predict the majority of first-order patterns of biodiversity on our planet—namely, ambient temperature and community size. Temperature primarily affects the rate of community turnover and the speed of evolution, while community size determines the number of individuals on which evolutionary processes can act. To use a chemical analogy, a community of individuals may be equivalent to the number of molecules reacting with each other in a test tube: the more molecules are interacting, and the higher the temperature, the higher the probability that new compounds will form purely by chance. Or, by similar analogy, life emerged on a primitive Earth almost 4 billion years ago (Tashiro et al. 2017), presumably through random chemical reactions at elevated temperature, whereas more complex organic compounds formed from simpler inorganic precursors (Miller and Orgel 1974). This analogy also helps us to highlight a significant weakness of the theory—namely, that the step from random mutations and molecular evolution to speciation rate is not mechanistically resolved, much like the step from randomly formed organic compounds to functioning cells is not clear. Even though our theory is not fully mechanistic due to this gap in understanding, it is encouraging that with an assumed link between temperature and speciation rate, and realistic spatial structure, the global patterns of biodiversity are captured surprisingly well (chapter 5).

The empirical and theoretical synthesis offered here may therefore help in some way to resolve a long-standing debate on the causes and mechanisms that underlie observed patterns of species richness at large scales (Rohde 1992; Gaston 2000; Gotelli et al. 2009; Brown 2014). From a theoretical perspective, this is significant because it sheds light on some of the processes that may drive first-order biogeographic patterns on our planet. From an applied perspective, this knowledge has relevance for understanding and conserving global biodiversity in the face of human impacts and global environmental change (chapter 6). From a methods perspective, our theory provides a flexible "modeling toolbox" to test alternative hypotheses on the relative importance of changes in environmental factors (for example, temperature, productivity), metacommunity processes (for example, dispersal), ecological attributes (for example, niche width), or even sampling effects (for example, the mid-domain effect) in driving the spatial organization of biodiversity. Of course, this toolbox could be extended, as there are many other processes (for example, the effects of large-scale disturbances, or habitat complexity) that could be specified and tested within this modeling framework. Ultimately, we hope that others will take our approach as a starting point and push it further, by improving the realism, confronting it with more data, specifying missing mechanisms, or invoking alternative hypotheses and approaches. We consider our theoretical and modeling framework as complementary to the statistical modeling more commonly used in macroecology and biodiversity science (Gotelli et al. 2009). Notwithstanding some outstanding questions about specific mechanisms, we hope that our work will help spur the development and growth of *mechanistic macroecology*, a more process-oriented approach to understanding large-scale patterns of biological variation (Purves et al. 2013; Harfoot et al. 2014).

7.2. ECOLOGICAL THEORY

Our work is founded on the Unified Neutral Theory of Biodiversity and Biogeography (Hubbell 2001), which itself built on the Theory of Island Biogeography (MacArthur and Wilson 1967). Both are neutral models, in which no inherent ecological differences among species need to be invoked to explain observed patterns of species richness. While MacArthur and Wilson's theory focused on local ecological processes, specifically local community size and dispersal, Hubbell extended this to include the process of evolution—namely, by incorporating a term for speciation. Likewise, our theory combines ecological and evolutionary processes—namely, speciation-extinction and dispersal-colonization dynamics (see fig. 7.1). We then translated this theory into a spatially explicit model by adapting Hubbell's neutral metacommunity approach to a global scale, and then adding empirically documented constraints on both evolutionary (such as temperature-driven metabolic rates) and ecological (thermal niches, barriers to dispersal) processes. In its full implementation, the model relaxes the assumption of neutrality, such that species can differ in their realized speciation rate (depending on the observed temperature) and environmental tolerance (depending on their thermal niche).

Despite including a variety of key processes, the model is still very simple, involving a maximum of four free parameters—namely, local community size (J) speciation rate (v), dispersal rate (m), and thermal niche width (w)—and a limited number of constants such as metacommunity size (J_M), and the Boltzmann and activation energy constants specified by metabolic theory. Any of the three parameters can be muted or amplified, allowing exploration of the relative importance of different processes in driving community composition and diversity. Similarly, the ecological and evolutionary mechanisms in our model can be individually switched on or off. In our explorations, we consistently found that speciation rate and community size were the most important variables in explaining contrasts in species richness. This is similar to results obtained by Hubbell (2001), who found that one could predict a wide range of local community properties from speciation rate and metacommunity size, which he combined into a composite parameter, the fundamental biodiversity number θ. Yet neutral theory so far has not yet been able to predict large-scale patterns of biodiversity, because it did not specify the effects of environmental drivers such as temperature or productivity on speciation rate and community size, respectively. This was accomplished here by an integration of neutral and metabolic theory.

Surveying a wide range of models and theories in community ecology, Vellend (2010) highlighted that patterns in the composition and diversity of species are fundamentally influenced by only four processes: selection, drift, speciation, and dispersal (see fig. 7.1B). Selection, in his notation, represents deterministic fitness differences among species, drift represents stochastic changes in species abundance, speciation creates new species, and dispersal is the movement of organisms across space (Vellend 2010). He suggested that all models in community ecology could be understood through the lens of these processes, but that none actually included all four of them. Vellend (2010) then conceptualized a general theory of community dynamics, in which species are added to communities via speciation and dispersal, and the relative abundances of these species are then shaped by drift and selection, as well as ongoing dispersal, to drive community dynamics (see fig. 7.1B). This theory was later implemented in a neutral two-species community model that served to illustrate some of the underlying processes (Vellend 2016), and hypotheses generated by the theory were discussed and compared against empirical data. Generally, there are conceptual similarities between Vellend (2016) and our volume, the major differences being the inclusion of metabolic theory, and the focus on large-scale biogeographic patterns herein (see fig. 7.1A). We believe that our theory is the first of global biodiversity to integrate all of Velland's key processes, re-create them in a model, and translate their effects from local communities to the global scale. Our theory might also be considered "efficient" *sensu* Marquet et al. (2014), as it is grounded in first principles and mathematical expressions, makes few assumptions, generates a large number of predictions per free parameter, and enables the testing of model predictions against empirical data, as demonstrated in chapter 5. We are, of course, aware of its limitations, and of the trade-offs between simplicity and realism. Thus, we make no claims of providing a final answer to some of these questions; this is merely a first attempt at reconciling some fundamental processes in a relatively coarse, abstract, and simple approach.

7.3. A NICHE FOR NEUTRALITY?

Our theory might perhaps help to inform an ongoing debate on the importance of neutral (independent of species traits) versus niche (species-specific adaptations) processes in shaping macroecological patterns. We find that on the large scales that we consider here, thermal niches do not emerge as a major factor in the generation of global species richness patterns, though they do have a clear influence on both species composition and community turnover (chapter 4). Indeed, if we were concerned with the *identity* of the species that make up our local communities, or with metrics that took such identities into account, then niches would clearly play a large role in constraining those. However, reasonably realistic latitudinal distributions along temperature gradients consistently emerged in a neutral-metabolic metacommunity whether niches were assumed or not. Our model results suggested that the average species will not establish a large range, not necessarily because of limited niche width, but because of the difficulty of colonizing new habitats and establishing a larger range across successive neighboring communities (Note that the model assumption of zero-sum ecological dynamics is important in this context.) The model results indicated that it is far more likely for a new species to become extinct than it is for it to grow in abundance and to spread to other regions. Thermal niches did significantly constrain realized niches only in cases where we assumed every disturbance led to a metacommunity dispersal event (equivalent to unconstrained dispersal in an open metacommunity). This might be a reasonable scenario in the pelagic realm, where species are very mobile and may disperse widely (Block et al. 2011; Beaugrand 2014), showing wider geographic and thermal niches than their coastal counterparts (see fig. 4.13). Yet a high dispersal scenario with niches did not provide an improved prediction of the observed data, even for pelagic taxa (see fig. 5.8). Of course, thermal niches are an important and heavily conserved feature of individual biology (Peterson et al. 1999). They are clearly important for species' realized distributions under a given set of environmental constraints (Sunday et al. 2011) and strongly influence local community composition and its response to changing climates (Sunday et al. 2012; Pinsky et al. 2013; Beaugrand et al. 2015). Moreover, the niche concept invokes a multidimensional axis, along which a species' ecological operating space is defined (Hutchinson 1957). Such multidimensional niches have not been considered in the framework of this theory, although this could be a possible extension for further work.

The niche versus neutrality debate highlights a key difference in terms of conceptualizing ecological processes. Here, we are emphasizing the perspective of (meta-)communities rather than that of individual species. That is, we are looking at aggregate community properties, specifically the total number of individuals and species, without being necessarily interested in their identity. For any given species, dispersal and niche limitations may both be constraining, as shown empirically for plankton communities (Chust et al. 2012), for example. But from the perspective of a whole community, and when integrating over evolutionary timescales, turnover in species identities may be less important for the long-term dynamics in species richness at large sales, which may be driven primarily by changes in speciation rate and community size. Interestingly, these conclusions are not unlike those of Vellend (2016), based on his more local theory of ecological communities. Therefore, the niche versus neutrality debate might perhaps move towards reconciliation by taking into account multiple viewpoints (species or community), timescales (ecological versus evolutionary), and metrics (focusing on species identities or broader properties).

7.4. SPATIAL SCALE

One of the fundamental premises of this book was the focus on large scales, specifically on regional to global patterns and drivers of biodiversity. Patterns and drivers of biodiversity do change across scales, however, likely becoming more idiosyncratic at smaller grain sizes, as local variations in ecological interactions and processes become more important in selecting from the regional species pool.

For land vertebrates, for example, it has been shown that the predictability of species richness patterns decreases sharply at smaller scales and that temperature and moisture as major drivers of species richness are replaced by other factors, such as productivity and environmental heterogeneity (Belmaker and Jetz 2011). This means that our theory at present captures some global-scale processes at coarse resolution quite well, but would need to be modified to apply at smaller scales.

One advantage of our metacommunity model is that it is easily scalable. While we considered only a tractable number of local communities (<1000), each of which was unrealistically small, there is no reason, other than computational cost, why this could not be scaled up to include many thousands of local communities, each of which could include large numbers of individuals. Our limited explorations in chapter 4 suggested that increasing the number of individuals in each local community does not qualitatively change the observed patterns (see also Tittensor and Worm 2016). Fitting a conceptually similar model to local or regional data and using realistic dispersal constraints would allow a closer look at processes operating at a variety of scales. As such, this approach could perhaps help to illuminate and test hypotheses about cross-scale variation in the processes and drivers of species richness.

7.5. ECOLOGICAL VERSUS EVOLUTIONARY TIME

Several authors have recently called upon ecologists to better integrate ecological and evolutionary processes in the study of communities and ecosystems (Fritz et al. 2013; Mannion et al. 2014; Weigelt et al. 2016). Our theory attempts such integration in an effort to explain global biodiversity patterns (see fig. 7.1). Within this framework, and over evolutionary timescales, assumed thermal effects on speciation emerge as the most general process that determines the speed at which new species evolve. At ecological timescales, habitat area and productivity adjust the number of individuals that can be supported, and thereby affect the number of species in a community. This also relates to spatial scale: at local scales, contemporary ecological drivers appear important in determining community size and selecting from the regional species pool. At larger scales, the size of the species pool, as determined by evolutionary drivers, becomes more important (Vellend 2016). This change from ecological to evolutionary drivers with increasing spatial scale might help to explain the empirically observed shift from productivity and habitat at smaller scales to temperature as the major environmental predictor at larger scales (see fig. 3.2 and data in Belmaker and Jetz 2011).

Biogeographically, the global distribution of thermal energy from a hot equator to cold poles is determined by the angle of the planet toward the sun, which is approximately stable over evolutionary time, though with variation related to regular Milankovitch cycles (Bennett 1990). As such, it makes sense that some latitudinal biodiversity gradient has persisted through time (Stehli et al. 1969; Yasuhara et al. 2012; Mannion et al. 2014). In contrast to the relative stability of global temperature gradients, changes in habitat area and productivity have been more dynamic. Indeed, these factors can show dramatic changes over both space and time—for example, through tectonic processes—and hence lead to varying patterns of species diversity such as "hopping hotspots" (Renema et al. 2008). Likewise, historic patterns of productivity and habitat area can leave a clear imprint on present-day patterns of diversity, shown both on continents (Jetz and Fine 2012) and on islands (Weigelt et al. 2016). We are as yet unable to include such temporal process variation within our coalescence model, although these could, at least conceptually, be implemented in our forward-simulation approach.

We suggest, however, that such temporal-spatial variation in habitat area and productivity is unlikely to override or reverse the long-term evolutionary signal of an equator-to-pole temperature gradient. Maybe the evolutionary primacy of temperature is one reason why this driver explains most of the variation in species richness across taxa, with habitat explaining a smaller proportion for most species groups (for example, Tittensor et al. 2010). Somewhat surprisingly, this also applies to mammals and birds, which operate at near-constant body temperature, yet are affected at least to some extent by the effects of ambient temperature (chapters 3 and 5), either directly or indirectly. While the precise mechanism by which ambient temperature drives evolutionary rates in these species groups is not necessarily clear, there is substantial empirical evidence for faster evolutionary rates in the tropics, both for ectotherms and for endotherms (Gillman et al. 2009; Gillman and Wright 2014), as well as higher origination rates in warmer environments (Jablonski et al. 2006), and during warmer periods over Earth's history (Mayhew et al. 2012).

7.6. APPLICATIONS

A clear understanding of biodiversity patterns, drivers, and mechanisms is required for global conservation planning, particularly in the face of rapid environmental change. A global synthesis of species richness such as presented in chapter 2 may help with prioritization of areas of elevated biodiversity, threat, and vulnerability, though perhaps could be repeated at a finer resolution where possible. The empirical outcome that patterns of biodiversity are highly correlated across species groups, and almost universally linked to temperature, habitat area, and productivity, also means that we need to pay particular attention to these factors in conservation biology. Temperature, habitat availability, and productivity are perhaps the three aspects of our planet most strongly influenced by people, and all three have been changing rapidly since the industrial revolution (Lotze and McClenachan 2013; Waters et al. 2016). Clearly, a permanently changed biosphere will produce novel communities and altered patterns of species richness in the short term, but also affect the course of evolution in the long term. Of course, these are not fundamentally new insights, as hotspots of richness, temperature change, and habitat change have been previously suggested as global conservation priorities (Tittensor et al. 2010; Selig et al. 2014; Beaugrand et al. 2015). What is new, however, is that we can now potentially integrate our understanding of the ecological and evolutionary processes that may generate biodiversity patterns in the first place with our knowledge of the major human impacts that compromise biodiversity. This integrated understanding may one day help us to predict future changes from a more mechanistic basis, and taking ecological and well as evolutionary dynamics into account. In this context, however, it is important to recall that we are not, within our framework, able to predict any real-world changes in species composition or identity, due to neutral model assumptions. We focus on aggregate variables concerning communities of competing organisms at the same trophic level, though models for different trophic groups could potentially be "stacked" and interact with each other. These limitations do of course restrict applications in conservation biology, which often maintains a focus on specific threatened species or species groups. However, there is also an increasing focus on whole-ecosystem approaches, and a growing interest in process-based modeling to be utilized for global conservation planning (Purves et al. 2013; Harfoot et al. 2014).

7.7. LIMITATIONS

We prefer our theory of global biodiversity to be interpreted as a series of hypotheses that are established through empirical and theoretical synthesis, and then tested and explored using a simple model that generates patterns from first principles based on putative mechanisms. The approach will continue to be confronted by new and additional data, and we remain excited to find out how well, or poorly, it may meet those challenges. Indeed, often it is the times when a model does not fit well that are the most informative and interesting (Hilborn and Mangel 1997), and it is in this spirit that we present our theory and recognize its many limitations, the most important of which we list here.

The neutral model, as it was originally developed, was geared toward organisms of a similar trophic level competing for similar sets of resources, hence resulting in zero-sum dynamics (Hubbell 2001). Thus, our approach does not explicitly include trophic interactions such as predation or competition (though mortality events may implicitly be caused by either). Other effects that are missing, but are known to be important, include the effects of habitat complexity and body size. We propose that these are reasonable assumptions and limitations for a first iteration of our model framework applied to very large scales.

Fully neutral models also have recognized limitations or assumptions in terms of their simplified representation of speciation, assumptions of species equivalence, evolutionary rates, and variable fit to empirical patterns (Bell 2001; Gravel et al. 2006; Rosindell et al. 2011; Marquet et al. 2014). In part, such limitations have provided incentive for us to relax the assumption of neutrality and to include potentially important differences among species—for example, with respect to their thermal tolerances. At the same time, we agree with other commentators who have pointed out that grossly simplified models can be very useful tools even when some of their assumptions are "wrong" (as with any model), by helping to inform us which processes may be important (Marquet et al. 2014).

Other inherent limitations of our model framework relate to abstractions in time and space. In particular, we do not attempt to define a specific temporal period that corresponds to each model time step. This is especially true for our modified version of Hubbell's model, which features probabilistic selection of single local communities for disturbance, rather than simultaneous or probabilistic disturbance of all communities. Our representation of time and time-specific rates is thus abstracted, and not easily comparable to empirically measured time steps or rates. Our model is also running on a coarse spatial grid, but using very small local communities, mostly for reasons of computational efficiency. At such scales, and particularly when fitting to a global grid, the meaning of a model "individual" does not equate to an individual in the real world, but more likely to a population. While we have tested both finer resolution metacommunities and larger local communities, we recognize that neither of these scenarios is strictly "realistic." The same goes for our parameter values of speciation rate and dispersal rate. Because both are scaled relative to a theoretical time step, of unspecified length, these rates do not yet have solid empirical grounding, but serve as tools for exploring fundamental processes. Again, we emphasize that this is a theoretical model that strives more for simplicity, tractability, and understanding rather than realism.

In summary, the degree of valid inference is limited by the simple nature of the mechanistic model that we derived. The emphasis here is on first-order patterns and processes, and model assumptions remain to be scrutinized or challenged by future work. Nonetheless, at present, no equivalent model exists, so we understand this as a "first step" rather than a finish line. The model in its simplicity and generality can be extended in many ways—for example, to include other factors such as environmental variability and large-scale disturbances, and test alternative hypotheses related to these and other processes.

7.8. FINAL OUTLOOK

How does species richness vary across the planet, and can we predict some of the observed patterns from simple ecological theory and first principles? In this volume, we have developed a synthetic empirical understanding of global patterns in species richness and developed a theory that can predict first-order patterns of species richness on land and in the sea. Much work remains to be done in testing, revising, and extending this theory. In closing, we point out some of the further research questions we see emerging from this work.

Clearly, more detailed analysis should be conducted to explain spatial patterns of biodiversity, especially at finer spatial grains. Given the greater availability of data on land and recent integration in global databases, it may be feasible to understand patterns and mechanisms in finer detail than in the ocean. This might also provide new insights about changing drivers and correlates of diversity across scales (Belmaker and Jetz 2011) and the factors that shape the exchange between local, regional, and global species pools (Vellend 2010, 2016).

Another open question is how the speed of evolution in endotherms and ectotherms is related to environmental conditions. How do gradients in temperature or productivity affect these groups similarly or differently? It is interesting, for example, that vascular plants (see fig. 2.8) and mammals on land (see fig. 2.9) show very similar geographic patterns of biodiversity, whereas those patterns diverge in the oceans for the same species groups (compare seagrasses and mangroves [see fig. 2.3] to cetaceans and pinnipeds [see figs. 2.4 and 2.6]). Yet in both environments, these groups strongly relate to gradients in surface temperature, albeit in different ways (see fig. 3.4). Clearly, the mechanistic basis of these relationships is not well resolved, especially for endotherms, but is an area of active research (Gillman and Wright 2014).

More broadly, our model could also be used to explore other macroecological patterns and biodiversity attributes in greater detail, such as species-area or range-abundance relationships. We have touched on this in our investigations of Rapoport's rule (see fig. 4.15), but clearly there is much more that can be done. Our strong focus on global richness has not allowed for much detailed exploration of other facets of biodiversity, but such work could potentially be accomplished with the tools and data at hand.

It would also be fascinating to examine past patterns of species richness under previous environmental regimes, and to test some of the ideas that are emerging in paleontology about dynamic changes in biodiversity throughout Earth's history (Renema et al. 2008; Mannion et al. 2012; Mannion et al. 2014). Can we reconstruct diversity patterns at various points in the past? As fundamental ecological laws on thermodynamics and resource availability should apply to all time periods, such an approach would generate "temporal replicates" to examine the ideas presented in this volume.

Likewise, potential applications in astrobiology would be fascinating to explore. Would temperature and habitat also have primacy in shaping richness patterns of life forms on other planets, which may well be based on alternative molecular building blocks and levels of organization? How do fundamental thermodynamic laws shape the course of evolution, no matter the circumstances? Clearly, there appears much room for exploration of the general principles that have shaped and continue to shape the distribution of life in our world, and possibly others. In looking back on what we have learned from the last 200 years of biodiversity research, we contend that life on Earth conforms to a set of universal laws that we must understand—and respect—in order to succeed in our quest for beneficial coexistence with those millions of other species that grace and define our planet.

References

Abele, L. G., and W. K. Patton. (1976). "The Size of Coral Heads and the Community Biology of Associated Decapod Crustaceans." *Journal of Biogeography* 3: 35–47.

Abele, L. G., and K. Walters. (1979). "The Stability-Time Hypothesis: Reevaluation of the Data." *American Naturalist* 114: 559–568.

Adler, P. B., J. HilleRisLambers, and J. M. Levine. (2007). "A Niche for Neutrality." *Ecology Letters* 10: 95–104.

Adler, P. B., E. W. Seabloom, E. T. Borer, H. Hillebrand, Y. Hautier, A. Hector, W. S. Harpole, L. R. O'Halloran, J. B. Grace, T. M. Anderson, J. D. Bakker et al. (2011). "Productivity Is a Poor Predictor of Plant Species Richness." *Science* 333, 1750–1753.

Agrawal, A. F., and A. D. Wang. (2008). "Increased Transmission of Mutations by Low-condition Females: Evidence for Condition-dependent DNA Repair." *PLoS Biology* 6: e30.

Airoldi, L., and M. W. Beck. (2007). "Loss, Status and Trends for Coastal Marine Habitats of Europe." In *Oceanography and Marine Biology*, vol. 45, ed. R. N. Gibson, R.J.A. Atkinson, and J.D.M. Gordon. Boca Raton, FL: CRC Press Taylor, and Francis Group, 345–405.

Algar, A. C., J. T. Kerr, and D. J. Currie. (2007). "A Test of Metabolic Theory as the Mechanism Underlying Broad-scale Species-richness Gradients." *Global Ecology and Biogeography* 16: 170–178.

Allen, A. P., J. H. Brown, and J. F. Gillooly. (2002). "Global Biodiversity, Biochemical Kinetics, and the Energetic-Equivalence Rule." *Science* 297: 1545–1548.

Allen, A. P., and J. F. Gillooly. (2006). "Assessing Latitudinal Gradients in Speciation Rates and Biodiversity at the Global Scale." *Ecology Letters* 9: 947–954.

Allen, A. P., J. F. Gillooly, V. M. Savage, and J. H. Brown. (2006). "Kinetic Effects of Temperature on Rates of Genetic Divergence and Speciation." *Proceedings of the National Academy of Sciences USA* 103: 9130–9135.

Allen, M. R., V. R. Barros, J. Broome, W. Cramer, R. Christ, J. A. Church, L. Clarke, Q. Dahe, P. Dasgupta, and N. K. Dubash. (2014). *IPCC Fifth Assessment Synthesis Report—Climate Change 2014 Synthesis Report*. Geneva: Intergovernmental Panel on Climate Change.

Alroy, J., M. Aberhan, D. J. Bottjer, M. Foote, F. T. Fursich, P. J. Harries, A.J.W. Hendy, S. M. Holland, L. C. Ivany, W. Kiessling, M. A. Kosnik et al. (2008). "Phanerozoic Trends in the Global Diversity of Marine Invertebrates." *Science* 321: 97–100.

Amend, A. S., T. A. Oliver, L. A. Amaral-Zettler, A. Boetius, J. A. Fuhrman, M. C. Horner-Devine, S. M. Huse, D.B.M. Welch, A. C. Martiny, and A. Ramette. (2013). "Macroecological Patterns of Marine Bacteria on a Global Scale." *Journal of Biogeography* 40: 800–811.

Angel, M. V. (1993). "Biodiversity of the Pelagic Ocean." *Conservation Biology* 7: 760–772.

————. (1997). "Pelagic Biodiversity." In *Marine Biodiversity*, ed. R.F.G. Ormond, J. D. Gage, and M. V. Angel. Cambridge, UK: Cambridge University Press, 35–69.

Araújo, M. B., F. Ferri-Yáñez, F. Bozinovic, P. A. Marquet, F. Valladares, and S. L. Chown. (2013). "Heat Freezes Niche Evolution." *Ecology Letters* 16: 1206–1219.

Arnason, U., A. Gullberg, A. Janke, M. Kullberg, N. Lehman, E. A. Petrov, and R. Väinölä. (2006). "Pinniped Phylogeny and a New Hypothesis for Their Origin and Dispersal. *Molecular Phylogenetics and Evolution* 41: 345–354.

Ausubel, J. H. (1999). "Toward a Census of Marine Life." *Oceanography* 12: 4–5.

Baker, S. M. (1910). "On the Causes of the Zoning of Brown Seaweeds on the Seashore." *New Phytologist* 9: 54–67.

Balsam, W. L., and K. W. Flessa. (1978). "Patterns of Planktonic Foraminiferal Abundance and Diversity in Surface Sediments of the Western North Atlantic." *Marine Micropaleontology* 3: 279–294.

Barnett, T. P., D. W. Pierce, and R. Schnur. (2001). "Detection of Anthropogenic Climate Change in the World's Oceans." *Science* 292: 270–273.

Barnosky, A. D., N. Matzke, S. Tomiya, G. O. Wogan, B. Swartz, T. B. Quental, C. Marshall, J. L. McGuire, E. L. Lindsey, and K. C. Maguire. (2011). "Has the Earth's Sixth Mass Extinction Already Arrived?" *Nature* 471: 51–57.

Barrett, R.D.H., A. Paccard, T. M. Healy, S. Bergek, P. M. Schulte, D. Schluter, and S. M. Rogers. (2010). "Rapid Evolution of Cold Tolerance in Stickleback." *Proceedings of the Royal Society B: Biological Sciences* 278: 239–246.

Bartlett, L. J., T. Newbold, D. W. Purves, D. P. Tittensor, and M.B.J. Harfoot. (2016). "Synergistic Impacts of Habitat Loss and Fragmentation on Model Ecosystems." *Proceedings of the Royal Society B: Biological Sciences* 283: e20161027.

Barton, A. D., S. Dutkiewicz , G. Flierl, J. Bragg, and M. J. Follows. (2010). "Patterns of Diversity in Marine Phytoplankton." *Science* 327: 1509–1511.

Baum, J. K., and R. A. Myers. (2004). "Shifting Baselines and the Decline of Pelagic Sharks in the Gulf of Mexico." *Ecology Letters* 7: 135–145.

Beaugrand, G. (2014). *Marine Biodiversity, Climatic Variability and Global Change.* Oxon, UK: Routledge.

Beaugrand, G., M. Edwards, V. Raybaud, E. Goberville, and R. R. Kirby. (2015). "Future Vulnerability of Marine Biodiversity Compared with Contemporary and Past Changes." *Nature Climate Change* 5: 695–701.

Beaugrand, G., F. Ibañez, J. A. Lindley, and P. C. Reid. (2002). "Diversity of Calanoid Copepods in the North Atlantic and Adjacent Seas: Species Associations and Biogeography." *Marine Ecology Progress Series* 232: 179–195.

Beaugrand, G., I. Rombouts, and R. R. Kirby. (2013). "Towards an Understanding of the Pattern of Biodiversity in the Oceans." *Global Ecology and Biogeography* 22: 440–449.

Behrenfeld, M. J., R. T. O'Malley, D. A. Siegel, C. R. McClain, J. L. Sarmiento, G. C. Feldman, A. J. Milligan, P. G. Falkowski, R. M. Letelier, and E. S. Boss. (2006). "Climate-driven Trends in Contemporary Ocean Productivity." *Nature* 444: 752–755.

Bell, G. (2001). "Neutral Macroecology." *Science* 293: 2413–2418.

Bellard, C., C. Bertelsmeier, P. Leadley, W. Thuiller, and F. Courchamp. (2012). "Impacts of Climate Change on the Future of Biodiversity." *Ecology Letters* 15: 365–377.

Bellwood, D. R., A. S. Hoey, J. L. Ackerman, and M. Depczynski. (2006). "Coral Bleaching, Reef Fish Community Phase Shifts and the Resilience of Coral Reefs." *Global Change Biology* 12: 1587–1594.

Bellwood, D. R., W. Renema, and B. R. Rosen. (2012). "Biodiversity Hotspots, Evolution and Coral Reef Biogeography: A Review." In *Biotic Evolution and Environmental Change in Southeast Asia*, ed. D. J. Gower, K. Johnson, J. Richardson, B. Rosen, L. Rüber, and S. Williams. Cambridge, UK: Cambridge University Press Cambridge, 216–242.

Belmaker, J., and W. Jetz. (2011). "Cross-scale Variation in Species Richness–Environment Associations." *Global Ecology and Biogeography* 20: 464–474.

Bennett, A. F., and R. E. Lenski. (1993). "Evolutionary Adaptation to Temperature II: Thermal Niches of Experimental Lines of *Escherichia Coli*." *Evolution* 47: 1–12.

Bennett, K. (1990). "Milankovitch Cycles and Their Effects on Species in Ecological and Evolutionary Time." *Paleobiology* 16: 11–21.

Benton, M. J., and R. J. Twitchett. (2003). "How to Kill (Almost) All Life: The End-Permian Extinction Event." *Trends in Ecology and Evol*ution 18: 358–365.

Berger, W. H., and F. L. Parker. (1970). "Diversity of Planktonic Foraminifera in Deep-sea Sediments." *Science* 168: 1345–1347.

Berke, S. K., D. Jablonski, A. Z. Krug, and J. W. Valentine. (2014). "Origination and Immigration Drive Latitudinal Gradients in Marine Functional Diversity." *PLoS ONE* 9: e101494.

Bleiweiss, R. (1998). "Slow Rate of Molecular Evolution in High-elevation Hummingbirds." *Proceedings of the National Academy of Sciences* 95: 612–616.

Block, B. A., I. D. Jonsen, S. J. Jorgensen, A. J. Winship, S. A. Shaffer, S. J. Bograd, E. L. Hazen, D. G. Foley, G. A. Breed, A. L. Harrison, J. E. Ganong et al. (2011). "Tracking Apex Marine Predator Movements in a Dynamic Ocean." *Nature* 475: 86–90.

Bouchet, P. (2006). "The Magnitude of Marine Biodiversity." In *The Exploration of Marine Biodiversity: Scientific and Technological Challenges*, ed. C. M. Duarte. Fundación BBVA, 33–64.

Bouchet, P., P. Lozouet, P. Maestrati, and V. Heros. (2002). "Assessing the Magnitude of Species Richness in Tropical Marine Environments: Exceptionally High Numbers of Molluscs at a New Caledonia Site." *Biological Journal of the Linnean Society* 75: 421–436.

Boyce, D., M. Lewis, and B. Worm. (2010). "Global Phytoplankton Decline over the Last Century." *Nature* 466: 591–596.

Boyce, D. G., M. Dowd, M. R. Lewis, and B. Worm. (2014). "Estimating Global Chlorophyll Changes over the Past Century." *Progress in Oceanography* 122: 163–173.

Boyce, D. G., D. P. Tittensor, and B. Worm. (2008). "Effects of Temperature on Global Patterns of Tuna and Billfish Richness." *Marine Ecology Progress Series* 355: 267–276.

Boyce, D. G., and B. Worm. (2015). "Patterns and Ecological Implications of Historical Marine Phytoplankton Change." *Marine Ecology Progress Series* 534: 251–272.

Brander, K. (2010). "Impacts of Climate Change on Fisheries." *Journal of Marine Systems* 79: 389–402.

Brandt, A., A. J. Gooday, S. N. Brandao, S. Brix, W. Brokeland, T. Cedhagen, M. Choudhury, N. Cornelius, B. Danis, I. De Mesel, R. J. Diaz et al. (2007). "First Insights into the Biodiversity and Biogeography of the Southern Ocean Deep Sea." *Nature* 447: 307–311.

Brey, T., M. Klages, C. Dahm, M. Gorny, J. Gutt, S. Hain, M. Stiller, W. E. Arntz, J.-W. Wägele, and A. Zimmermann. (1994). "Antarctic Benthic Diversity." *Nature* 368: 297–298.

Brill, R. W. (1994). "A Review of Temperature and Oxygen Tolerance Studies of Tunas Pertinent to Fisheries Oceanography, Movement Models, and Stock Assessments." *Fisheries Oceanography* 3: 204–216.

Britten, G. L., M. Dowd, and B. Worm. (2016). "Changing Recruitment Capacity in Global Fish Stocks." *Proceedings of the National Academy of Sciences* 113: 134–139.

Bromham, L., and M. Cardillo. (2003). "Testing the Link between the Latitudinal Gradient in Species Richness and Rates of Molecular Evolution." *Journal of Evolutionary Biology* 16: 200–207.

Brown, J. H. (1995). *Macroecology*. Chicago: University of Chicago Press.

———. (2014). Why Are There So Many Species in the Tropics?" *Journal of Biogeography* 41: 8–22.

Brown, J. H., J. F. Gillooly, A. P. Allen, V. M. Savage, and G. B. West. (2004). "Toward a Metabolic Theory of Ecology." *Ecology* 85: 1771–1789.

Bryden, H. L., H. R. Longworth, and S. A. Cunningham. (2005). "Slowing of the Atlantic Meridional Overturning Circulation at 25°N." *Nature* 438: 655–657.

Burrows, M. T., D. S. Schoeman, L. B. Buckley, P. Moore, E. S. Poloczanska, K. M. Brander, C. Brown, J. F. Bruno, C. M. Duarte, B. S. Halpern, J. Holding et al. (2011). "The Pace of Shifting Climate in Marine and Terrestrial Ecosystems." *Science* 334: 652–655.

Butchart, S. H., M. Walpole, B. Collen, A. van Strien, J. P. Scharlemann, R. E. Almond, J. E. Baillie, B. Bomhard, C. Brown, J. Bruno, K. E. Carpenter et al. (2010). "Global Biodiversity: Indicators of Recent Declines." *Science* 328: 1164–1168.

Butchart, S.H.M., A. J. Stattersfield, L. A. Bennun, S. M. Shutes, H. R. Akcakaya, J.E.M. Baillie, S. N. Stuart, C. Hilton-Taylor, and G. M. Mace. (2004). "Measuring Global Trends in the Status of Biodiversity: Red List Indices for Birds." *PLoS Biology* 2: 2294–2304.

Cairns, S. D. (2007). "Deep-water Corals: An Overview with Special Reference to Diversity and Distribution of Deep-water Scleractinian Corals." *Bulletin of Marine Science* 81: 311–322.

Cairns Jr., J. (2013). "Stress, Environmental." In *Encyclopedia of Biodiversity*, ed. S. Levin. Waltham, MA: Academic Press, 39–44.

Campbell, J. E., J. A. Berry, U. Seibt, Steven J. Smith, S. A. Montzka, T. Launois, S. Belviso, L. Bopp, and M. Laine. (2017). "Large Historical Growth in Global Terrestrial Gross Primary Production." *Nature* 544: 84–87.

Cermeño, P., C. de Vargas, F. Abrantes, and P. G. Falkowski. (2010). "Phytoplankton Biogeography and Community Stability in the Ocean." *PLoS ONE* 5: e10037.

Cermeño, P., and P. G. Falkowski. (2009). "Controls on Diatom Biogeography in the Ocean." *Science* 325: 1539–1541.

Chaudhary, C., H. Saeedi, and M. J. Costello. (2016). "Bimodality of Latitudinal Gradients in Marine Species Richness." *Trends in Ecology, and Evolution* 31: 670–676.

Cheung, W.W.L., R. D. Brodeur, T. A. Okey, and D. Pauly. (2015). "Projecting Future Changes in Distributions of Pelagic Fish Species of Northeast Pacific Shelf Seas." *Progress in Oceanography* 130: 19–31.

Cheung, W.W.L., V.W.Y. Lam, J. L. Sarmiento, K. Kearney, R. Watson, and D. Pauly. (2009). "Projecting Global Marine Biodiversity Impacts under Climate Change Scenarios." *Fish and Fisheries* 10: 235–251.

Cheung, W.W.L., R. Watson, and D. Pauly. (2013). "Signature of Ocean Warming in Global Fisheries Catch." *Nature* 497: 365–368.

Chisholm, R. A., and S. W. Pacala. (2010). "Niche and Neutral Models Predict Asymptotically Equivalent Species Abundance Distributions in High-diversity Ecological Communities." *Proceedings of the National Academy of Sciences* 107: 15821–15825.

Chust, G., X. Irigoien, J. Chave, and R. P. Harris. (2012). "Latitudinal Phytoplankton Distribution and the Neutral Theory of Biodiversity." *Global Ecology and Biogeography*; 22:531–543.

Clarke, A., and K. J. Gaston. (2006). "Climate, Energy and Diversity." *Proceedings of the Royal Society: Series B* 273: 2257–2266.

Colwell, R. K., and D. C. Lees. (2000). "The Mid-domain Effect: Geometric Constraints on the Geography of Species Richness." *Trends in Ecology and Evolution* 15: 70–76.

Comes, H. P., and J. W. Kadereit. (1998). "The Effect of Quaternary Climatic Changes on Plant Distribution and Evolution." *Trends in Plant Science* 3: 432–438.

Connell, J. H. (1978). "Diversity in Tropical Rain Forests and Coral Reefs." *Science* 199: 1302–1310.

Connell, J. H., and E. Orias. (1964). "The Ecological Regulation of Species Diversity." *American Naturalist* 98: 399–441.

Connor, E. F., A. C. Courtney, and J. M. Yoder. (2000). "Individuals-Area Relationships: The Relationship between Animal Population Density and Area." *Ecology* 81: 734–748.

Connor, E. F., and E. D. McCoy. (1979). "The Statistics and Biology of the Species-Area Relationship." *American Naturalist* 113: 791–833.

Corkrey, R., T. A. McMeekin, J. P. Bowman, D. A. Ratkowsky, J. Olley, and T. Ross. (2014). "Protein Thermodynamics Can Be Predicted Directly from Biological Growth Rates." *PLoS ONE* 9: e96100.

Cotton, C. M (1996). *Ethnobotany: Principles and Applications.* New York: John Wiley & Sons.

Cowen, R. K, C. B. Paris, and A. Srinivasan. (2006). "Scaling of Connectivity in Marine Populations." *Science* 311: 522–527.

Cowling, R. M., R. L. Pressey, M. Rouget, and A. T. Lombard. (2003). "A Conservation Plan for a Global Biodiversity Hotspot—the Cape Floristic Region, South Africa." *Biological Conservation* 112: 191–216.

Crame, J. A. (2000). "Evolution of Taxonomic Diversity Gradients in the Marine Realm: Evidence from the Composition of Recent Bivalve Faunas." *Paleobiology* 26: 188–214.

Crawley, M. J., and J. E. Harral. (2001). "Scale Dependence in Plant Biodiversity." *Science* 291: 864–868.

Culver, S. J., and M. A. Buzas. (2000). "Global Latitudinal Species Diversity Gradient in Deep-sea Benthic Foraminifera." *Deep Sea Research Part I: Oceanographic Research Papers* 47: 259–275.

Currie, D. J. (1991). "Energy and Large-scale Patterns of Animal and Plant Species Richness." *American Naturalist* 137: 27–49.

Currie, D. J., and J. T. Kerr. (2008). "Tests of the Mid-domain Hypothesis: A Review of the Evidence." *Ecological Monographs* 78: 3–18.

Darwin, C. (1859). *The Origin of Species.* Modern Library ed. New York: Random House.

Davies, R. G., U. M. Irlich, S. L. Chown, and K. J. Gaston. (2010). "Ambient, Productive and Wind Energy, and Ocean Extent Predict Global Species Richness of Procellariiform Seabirds." *Global Ecology and Biogeography* 19: 98–110.

Davison, J., M. Moora, M. Öpik, A. Adholeya, L. Ainsaar, A. Bâ, S. Burla, A. G. Diedhiou, I. Hiiesalu, T. Jairus, N. C. Johnson et al. (2015). "Global Assessment of

Arbuscular Mycorrhizal Fungus Diversity Reveals Very Low Endemism." *Science* 349: 970–973.

Devictor, V., D. Mouillot, C. Meynard, F. Jiguet, W. Thuiller, and N. Mouquet. (2010). "Spatial Mismatch and Congruence between Taxonomic, Phylogenetic and Functional Diversity: The Need for Integrative Conservation Strategies in a Changing World." *Ecology Letters* 13: 1030–1040.

De Vos, J. M., L. N. Joppa, J. L. Gittleman, P. R. Stephens, and S. L. Pimm. (2015). "Estimating the Normal Background Rate of Species Extinction." *Conservation Biology* 29: 452–462.

Dillon, M. E., G. Wang, and R. B. Huey. (2010). "Global Metabolic Impacts of Recent Climate Warming." *Nature* 467: 704–707.

Dirzo, R., H. S. Young, M. Galetti, G. Ceballos, N.J.B. Isaac, and B. Collen. (2014). "Defaunation in the Anthropocene." *Science* 345: 401–406.

Donner, S. D., W. J. Skirving, C. M. Little, M. Oppenheimer, and O.V.E. Hoegh-Guldberg. (2005). "Global Assessment of Coral Bleaching and Required Rates of Adaptation under Climate Change." *Global Change Biology* 11: 2251–2265.

Dormann, C. F., J. M. McPherson, M. B. Araújo, R. Bivand, J. Bolliger, G. Carl, R. G. Davies, A. Hirzel, W. Jetz, W. D. Kissling, I. Kühn et al. (2007). "Methods to Account for Spatial Autocorrelation in the Analysis of Species Distributional Data: A Review." *Ecography* 30: 609–628.

Dornelas, M., S. R. Connolly, and T. P. Hughes. (2006). "Coral Reef Diversity Refutes the Neutral Theory of Biodiversity." *Nature* 440: 80–82.

Dowle, E., M. Morgan-Richards, and S. Trewick. (2013). "Molecular Evolution and the Latitudinal Biodiversity Gradient." *Heredity* 110: 501–510.

Drakare, S., J. J. Lennon, and H. Hillebrand. (2006). "The Imprint of the Geographical, Evolutionary and Ecological Context on Species–Area Relationships." *Ecology Letters* 9: 215–227.

Duce, R. A., J. LaRoche, K. Altieri, K. R. Arrigo, A. R. Baker, D. G. Capone, S. Cornell, F. Dentener, J. Galloway, R. S. Ganeshram, R. J. Geider et al. (2008). "Impacts of Atmospheric Anthropogenic Nitrogen on the Open Ocean." *Science* 320: 893–897.

Edgar, G. J., R. D. Stuart-Smith, T. J. Willis, S. Kininmonth, S. C. Baker, S. Banks, N. S. Barrett, M. A. Becerro, A.T.F. Bernard, J. Berkhout, C. D. Buxton et al. (2014). "Global Conservation Outcomes Depend on Marine Protected Areas with Five Key Features." *Nature* 506: 216–220.

Effiom, E. O., G. Nuñez-Iturri, H. G. Smith, U. Ottosson, and O. Olsson. (2013). "Bushmeat Hunting Changes Regeneration of African Rainforests." Proceedings of the Royal Society of London B: Biological Sciences 280: e20130246.

Eggleton, P. (2000). "Global Patterns of Termite Diversity." In *Termites: Evolution, Sociality, Symbioses, Ecology*, ed Y. Abe, David Edward Bignell, and T. Higashi. New York: Springer, 25–51.

Ellis, E. C., E. C. Antill, and H. Kreft. (2012). "All Is Not Loss: Plant Biodiversity in the Anthropocene." *PLoS ONE* 7: e30535.

Ellis, E. C., K. Klein Goldewijk, S. Siebert, D. Lightman, and N. Ramankutty. (2010). "Anthropogenic Transformation of the Biomes, 1700 to 2000." *Global Ecology and Biogeography* 19: 589–606.

Elmqvist, T., C. Folke, M. Nyström, G. Peterson, J. Bengtsson, B. Walker, and J. Norberg. (2003). "Response Diversity, Ecosystem Change, and Resilience." *Frontiers in Ecology and the Environment* 1: 488–494.

Erb, K.-H., T. Fetzel, C. Plutzar, T Kastner, C. Lauk, A. Mayer, M. Niedertscheider, C. Körner, and H. Haberl. (2016). "Biomass Turnover Time in Terrestrial Ecosystems Halved by Land Use." *Nature Geoscience* 9: 674–678.

Eschmeyer, W. N., R. Fricke, J. D Fong, and D. A Polack. (2010). "Marine Fish Diversity: History of Knowledge and Discovery (Pisces)." *Zootaxa* 2525: 19–50.

Estes, J. A., J. Terborgh, J. S. Brashares, M. E. Power, J. Berger, W. J. Bond, S. R. Carpenter, T. E. Essington, R. D. Holt, J.B.C. Jackson, R. J. Marquis et al. (2011). "Trophic Downgrading of Planet Earth." *Science* 333: 301–306.

Etnoyer, P., D. Canny, B. Mate, and L. Morgan. (2004). "Persistent Pelagic Habitats in the Baja California to Bering Sea (B2B) Ecoregion." *Oceanography* 17: 90–101.

Evans, K. L., P. H. Warren, and K. J. Gaston. (2005). "Species–Energy Relationships at the Macroecological Scale: A Review of the Mechanisms." *Biological Reviews* 80: 1–25.

Fahrig, L. (2003). "Effects of Habitat Fragmentation on Biodiversity." *Annual Reviews in Ecology and Systematics* 34: 487–515.

Falkowski, P. (2012). "The Power of Plankton." *Nature* 483: S17–S20.

Felsenstein, J. (2004). *Inferring Phylogenies*. Sunderland, MA: Sinauer Associates.

Fernandez, M., A. Astorga, S. A. Navarrete, C. Valdovinos, and P. A. Marquet. (2009). "Deconstructing Latitudinal Species Richness Patterns in the Ocean: Does Larval Development Hold the Clue." *Ecology Letters* 12: 601–611.

Ferraroli, S., J.-Y. Georges, P. Gaspar, and Y. L. Maho. (2004). "Where Leatherback Turtles Meet Fisheries." *Nature* 429: 521–522.

Field, C. B., M. J. Behrenfeld, J. T. Randerson, and P. Falkowski. (1998). "Primary Production of the Biosphere: Integrating Terrestrial and Oceanic Components." *Science* 281: 237–240.

Fierer, N., and R. B. Jackson. (2006). "The Diversity and Biogeography of Soil Bacterial Communities." *Proceedings of the National Academy of Sciences* 103: 626–631.

Fischer, A. G. (1960). "Latitudinal Variations in Organic Diversity." *Evolution* 14: 64–81.

Floeder, S., and U. Sommer. (2000). "An Experimental Test of the Intermediate Disturbance Hypothesis Using Large Limnetic Enclosures." *Verhandlungen des Internationalen Verein Limnologie* 27: 2892–2893.

Franklin, J., F. W. Davis, M. Ikegami, A. D. Syphard, L. E. Flint, A. L. Flint, and L. Hannah. (2013). "Modeling Plant Species Distributions under Future Climates: How Fine Scale Do Climate Projections Need to Be?" *Global Change Biology* 19: 473–483.

Fritz, S. A., J. Schnitzler, J. T. Eronen, C. Hof , K. Böhning-Gaese, and C. H. Graham. (2013). "Diversity in Time and Space: Wanted Dead and Alive." *Trends in Ecology, and Evolution* 28: 509–516.

Fuhrman, J. A., J. A. Steele, I. Hewson, M. S. Schwalbach, M. V. Brown, J. L. Green, and J. H. Brown. (2008). "A Latitudinal Diversity Gradient in Planktonic Marine Bacteria." *Proceedings of the National Academy of Sciences USA* 105: 7774–7778.

Fulton, E. A., A.D.M. Smith, and A. E. Punt. (2005). "Which Ecological Indicators Can Robustly Detect Effects of Fishing?" *ICES Journal of Marine Science* 62: 540–551.

Gaston, K. J. (2000). "Global Patterns in Biodiversity." *Nature* 405: 220–227.

Gillman, L. N., D. J. Keeling, H. A. Ross, and S. D. Wright. (2009). "Latitude, Elevation and the Tempo of Molecular Evolution in Mammals." *Proceedings of the Royal Society of London B: Biological Sciences* 276: 3353–3359.

Gillman, L. N., and S. D. Wright. (2014). "Species Richness and Evolutionary Speed: The Influence of Temperature, Water and Area." *Journal of Biogeography* 41: 39–51.

Gillooly, J. F., A. P. Allen, G. B. West, and J. H. Brown. (2005). "The Rate of DNA Evolution: Effects of Body Size and Temperature on the Molecular Clock." *Proceedings of the National Academy of Sciences USA* 102: 140–145.

Gillooly, J. F., J. H. Brown, G. B. West, V. M. Savage, and E. L. Charnov. (2001). "Effects of Size and Temperature on Metabolic Rate." *Science* 293: 2248–2251.

Godfray, H.C.J., J. R. Beddington, I. R. Crute, L. Haddad, D. Lawrence, J. F. Muir, J. Pretty, S. Robinson, S. M. Thomas, and C. Toulmin. (2010). "Food Security: The Challenge of Feeding 9 Billion People." *Science* 327: 812–818.

Gotelli, N. J., M. J. Anderson, H. T. Arita, A. Chao, R. K. Colwell, S. R. Connolly, D. J. Currie, R. R. Dunn, G. R. Graves, and J. L. Green. (2009). "Patterns and Causes of Species Richness: A General Simulation Model for Macroecology." *Ecology Letters* 12: 873–886.

Grassle, J. F., and N. J. Maciolek. (1992). "Deep-sea Species Richness: Regional and Local Diversity Estimates from Quantitative Bottom Samples." *American Naturalist* 139: 313–341.

Grassle, J. F., and H. L. Sanders. (1973). "Life Histories and the Role of Disturbance." *Deep Sea Research and Oceanographic Abstracts* 20: 643–659.

Grassle, J. F., and K. I. Stocks. (1999). "A Global Ocean Biogeographic Information System (OBIS) for the Census of Marine Life." *Oceanography* 12: 12–14.

Gravel, D., C. D. Canham, M. Beaudet, and C. Messier. (2006). "Reconciling Niche and Neutrality: The Continuum Hypothesis." *Ecology Letters* 9: 399–409.

Gravel, D., F. Guichard, and M. E. Hochberg. (2011). "Species Coexistence in a Variable World." *Ecology Letters* 14: 828–839.

Gregg, W. W., M. E. Conkright, Paul Ginoux, J. E. O'Reilly, and Nancy W. Casey. (2003). "Ocean Primary Production and Climate: Global Decadal Changes." *Geophysical Research Letters* 30: 1909–1813.

Grenyer, R., C.D.L. Orme, S. F. Jackson, G. H. Thomas, R. G. Davies, T. J. Davies, K. E. Jones, V. A. Olson, R. S. Ridgely, P. C. Rasmussen, T.-S. Ding et al. (2006). "Global Distribution and Conservation of Rare and Threatened Vertebrates." *Nature* 444: 93–96.

Grime, J. P. (1973). "Competitive Exclusion in Herbaceous Vegetation." *Nature* 242: 344–347.

Grimm, V., and C. Wissel. (1997). "Babel, or the Ecological Stability Discussions: An Inventory and Analysis of Terminology and a Guide for Avoiding Confusion." *Oecologia* 109: 323–334.

Haedrich, R. L., and G. T. Rowe. (1977). "Megafaunal Biomass in the Deep Sea." *Nature* 269: 141–142.

Halpern, B., M. Frazier, J. Potapenko, K. Casey, K. Koenig, et al. 2015. *Cumulative Human Impacts: Supplementary Data.* Knowledge Network for Biocomplexity. doi:10.5063/F19Z92TW.

Halpern, B. S., S. Walbridge, K. A. Selkoe, C. V. Kappel, F. Micheli, C. D'Agrosa, J. F. Bruno, K. S. Casey, C. Ebert, H. E. Fox, R. Fujita et al. (2008). "A Global Map of Human Impact on Marine Ecosystems." *Science* 319: 948–952.

Haney, J. C. (1986). "Seabird Aggregation at Gulf Stream Frontal Eddies." *Marine Ecology Progress Series* 28: 279–285.

Hansen, J., R. Ruedy, M. Sato, and K. Lo. (2010). "Global Surface Temperature Change." *Reviews of Geophysics* 48; doi:10.1029/2010RG000345.

Harfoot, M. B., T. Newbold, D. P. Tittensor, S. Emmott, J. Hutton, V. Lyutsarev, M. J. Smith, J. P. Scharlemann, and D. W. Purves. (2014). "Emergent Global Patterns of Ecosystem Structure and Function from a Mechanistic General Ecosystem Model." *PLoS Biology* 12: e1001841.

Harley, C.D.G., K. M. Anderson, K. W. Demes, J. P. Jorve, R. L. Kordas, T. A. Coyle, and M. H. Graham. (2012). "Effects of Climate Change on Global Seaweed Communities." *Journal of Phycology* 48: 1064–1078.

Harvey, P. H., and M. Pagel. (1991). *The Comparative Method in Evolutionary Biology.* Oxford, UK: Oxford University Press.

Hawkins, B. A., F. S. Albuquerque, M. B. Araujo, J. Beck, L. M. Bini, F. J. Cabrero-Sanudo, I. Castro-Parga, J.A.F. Diniz-Filho, D. Ferrer-Castan, R. Field, J. F. Gómez et al. (2007). "A Global Evaluation of Metabolic Theory as an Explanation for Terrestrial Species Richness Gradients." *Ecology* 88: 1877–1888.

Hawkins, B. A., J.A.F. Diniz-Filho, C. A. Jaramillo, and S. A. Soeller. (2006). "Post-Eocene Climate Change, Niche Conservatism, and the Latitudinal Diversity Gradient of New World Birds." *Journal of Biogeography* 33: 770–780.

Hawkins, B. A., and J.A.F. Diniz-Filho. (2002). "The Mid-domain Effect Cannot Explain the Diversity Gradient of Nearctic Birds." *Global Ecology and Biogeography* 11: 419–426.

Hawkins, B. A., R. Field, H. V. Cornell, D. J. Currie, J.-F. Guégan, D. M. Kaufman, J. T. Kerr, G. G. Mittelbach, T. Oberdorff, E. M. O'Brien, E. E. Porter, and J.R.G. Turner. (2003). "Energy, Water, and Broad-scale Geographic Patterns of Species Richness." *Ecology* 84: 3105–3117.

Hays, G. C., A. J. Richardson, and C. Robinson. (2005). "Climate Change and Marine Plankton." *Trends in Ecology and Evolution* 20: 337–344.

Hazen, E. L., R. M. Suryan, J. A. Santora, S. J. Bograd, Y. Watanuki, and R. P. Wilson. (2013). "Scales and Mechanisms of Marine Hotspot Formation." *Marine Ecology Progress Series* 487: 177–183.

He, F., and S. P. Hubbell. (2011). "Species-Area Relationships Always Overestimate Extinction Rates from Habitat Loss." *Nature* 473: 368–371.

Henderson, P. A. (2007). "Discrete and Continuous Change in the Fish Community of the Bristol Channel in Response to Climate Change." *Journal of the Marine Biological Association of the United Kingdom* 87: 589–598.

Hessler, R. R., and H. L. Sanders. (1967). "Faunal Diversity in the Deep Sea." *Deep Sea Research* 14: 65–78.

Hewitt, G. M. (1996). "Some Genetic Consequences of Ice Ages, and Their Role in Divergence and Speciation." *Biological Journal of the Linnean Society* 58: 247–276.

Hiddink, J. G., and R. ter Hofstede. (2008). "Climate Induced Increases in Species Richness of Marine Fishes." *Global Change Biology* 14 (3): 453–460.

Hilborn, R., and M. Mangel. (1997). *The Ecological Detective: Confronting Models with Data.* Princeton, NJ: Princeton University Press.

Hillebrand, H. (2004). "On the Generality of the Latitudinal Diversity Gradient." *American Naturalist* 163: 192–211.

Hillebrand, H., F. Watermann, R. Karez, and U.-G. Berninger. (2001). "Differences in Species Richness Patterns between Unicellular and Multicellular Organisms." *Oecologia* 126: 114–124.

Hoegh-Guldberg, O. (1999). "Climate Change, Coral Bleaching and the Future of the World's Coral Reefs." *Marine and Freshwater Research* 50: 839–866.

Hoffmann, A. A., and C. M. Sgrò. (2011). "Climate Change and Evolutionary Adaptation." *Nature* 470: 479–485.

Holbrook, S. J., R. J. Schmitt, and J. S. Stephens Jr.. (1997). "Changes in an Assemblage of Temperate Reef Fishes Associated with a Climatic Shift." *Ecological Applications* 7: 1299–1310.

Honjo, S., and H. Okada. (1974). "Community Structure of Coccolithophores in the Photic Layer of the Mid-Pacific." *Micropaleontology* 20: 209–230.

Hooper, D. U., E. C. Adair, B. J. Cardinale, J.E.K. Byrnes, B. A. Hungate, K. L. Matulich, A. Gonzalez, J. E. Duffy, L. Gamfeldt, and M. I. O'Connor. (2012). "A Global Synthesis Reveals Biodiversity Loss As a Major Driver of Ecosystem Change." *Nature* 486: 105–108.

Hubbell, S. P. (2001). *The Unified Neutral Theory of Biodiversity and Biogeography.* Princeton, NJ: Princeton University Press.

Hughes, T. P., J. T. Kerry, M. Álvarez-Noriega, J. G. Álvarez-Romero, K. D. Anderson, A. H. Baird, R. C. Babcock, M. Beger, D. R. Bellwood, and R. Berkelmans. (2017). "Global Warming and Recurrent Mass Bleaching of Corals." *Nature* 543: 373–377.

Hurlbert, A. H., and W. Jetz. (2010). "More Than 'More Individuals': The Nonequivalence of Area and Energy in the Scaling of Species Richness." *American Naturalist* 176: E50–E65.

Huston, M. A. (1979). "A General Hypothesis of Species Diversity." *American Naturalist* 113: 81–101.

———. (1994). *Biological Diversity.* Cambridge, UK: Cambridge University Press.

Hutchings, P., and P. Saenger. (1987). *Ecology of Mangroves.* St. Lucia, Australia: Queensland University Press.

Hutchinson, G. E. (1957). "The Multivariate Niche." *Cold Spring Harbor Symposium Quantitative Biology* 22: 415–421.

———. (1959). "Homage to Santa Rosalia, or Why Are There So Many Kinds of Animals." *American Naturalist* 93: 145–159.

Imhoff, M. L., L. Bounoua, T. Ricketts, C. Loucks, R. Harris, and W. T. Lawrence. (2004). "Global Patterns in Human Consumption of Net Primary Production." *Nature* 429: 870–873.

International Union for the Conservation of Nature (IUCN). (2016). *Redlist Spatial Data.* Cambridge, UK: International Union for the Conservation of Nature Redlist Unit. Available at http://www.iucnredlist.org/technical-documents/spatial-data.

Jablonski, D., C. L. Belanger, S. K. Berke, S. Huang, A. Z. Krug, K. Roy, A. Tomasovych, and J. W. Valentine. (2013). "Out of the Tropics, but How? Fossils, Bridge Species, and Thermal Ranges in the Dynamics of the Marine Latitudinal Diversity Gradient." *Proceedings of the National Academy of Sciences* 110: 10487–10494.

Jablonski, D., K. Roy, and J. W. Valentine. (2006). "Out of the Tropics: Evolutionary Dynamics of the Latitudinal Diversity Gradient." *Science* 314: 102–106.

Jamieson, A. J., T. Malkocs, S. B. Piertney, T. Fujii, and Z. Zhang. (2017). "Bioaccumulation of Persistent Organic Pollutants in the Deepest Ocean Fauna." *Nature Ecology and Evolution* 1: e0051.

Jefferies, R. L., and J. L. Maron. (1997). "The Embarrassment of Riches: Athmospheric Deposition of Nitrogen and Community and Ecosystem Processes." *Trends in Ecology and Evolution* 12: 74–78.

Jetz, W., and P.V.A. Fine. (2012). "Global Gradients in Vertebrate Diversity Predicted by Historical Area-Productivity Dynamics and Contemporary Environment." *PLoS Biology* 10: e1001292.

Jetz, W., H. Kreft, G. Ceballos, and J. Mutke. (2009). "Global Associations between Terrestrial Producer and Vertebrate Consumer Diversity." *Proceedings of the Royal Society B: Biological Sciences* 276: 269–278.

Jetz, W., J. M. McPherson, and R. P. Guralnick. (2012a). "Integrating Biodiversity Distribution Knowledge: Toward a Global Map of Life." *Trends in Ecology, and Evolution* 27: 151–159.

Jetz, W., and C. Rahbek. (2002). "Geographic Range Size and Determinants of Avian Species Richness." *Science* 297: 1548–1551.

Jetz, W., G. Thomas, J. Joy, K. Hartmann, and A. Mooers. (2012b). "The Global Diversity of Birds in Space and Time." *Nature* 491: 444–448.

Johnston, I. A., and A. F. Bennett. (1996). *Animals and Temperature: Phenotypic and Evolutionary Adaptation*. Cambridge, UK: Cambridge University Press.

Kaiser, M. J., K. R. Clarke, H. Hinz, M.C.V. Austen, P. J. Somerfield, and I. Karakassis. (2006). "Global Analysis of Response and Recovery of Benthic Biota to Fishing." *Marine Ecology Progress Series* 311: 1–14.

Kareiva, P., and M. Marvier. (2003). "Conserving Biodiversity Coldspots." *American Scientist* 91: 344–351.

Kaschner, K., D. Tittensor, T. Ready, T. Gerrodette, and B. Worm. (2011). "Current and Future Patterns of Marine Mammal Biodiversity." *PLoS ONE* 6: e19653.

Kassen, R., A. Buckling, G. Bell, and P. B. Rainey. (2000). "Diversity Peaks at Intermediate Productivity in Laboratory Microcosms." *Nature* 406: 508–512.

Keith, S. A., A. P. Kerswell, and S. R. Connolly. (2014). "Global Diversity of Marine Macroalgae: Environmental Conditions Explain Less Variation in the Tropics." *Global Ecology and Biogeography*; 23: 517–529.

Kerswell, A. P. (2006). "Global Biodiversity Patterns of Benthic Marine Algae." *Ecology* 87: 2479–2488.

Kier, G., J. Mutke, E. Dinerstein, T. H. Ricketts, W. Küper, H. Kreft, and W. Barthlott. (2005). "Global Patterns of Plant Diversity and Floristic Knowledge." *Journal of Biogeography* 32: 1107–1116.

Kiessling, W., C. Simpson, B. Beck, H. Mewis, and J. M. Pandolfi. (2012). "Equatorial Decline of Reef Corals during the Last Pleistocene Interglacial." *Proceedings of the National Academy of Sciences* 109: 21378–21383.

Kleypas, J. (2015). "Invisible Barriers to Dispersal." *Science* 348: 1086–1087.

Kondoh, M. (2001). "Unifying the Relationships of Species Richness to Productivity and Disturbance." *Proceedings of the Royal Society of London B: Biological Sciences* 268: 269–271.

Kreft, H., and W. Jetz. (2007). "Global Patterns and Determinants of Vascular Plant Diversity." *Proceedings of the National Academy of Sciences USA* 104: 5925–5930.

Krug, A. Z., D. Jablonski, J. W. Valentine, and K. Roy. (2009). "Generation of Earth's First-order Biodiversity Pattern." *Astrobiology* 9: 113–124.

Ladau, J., T. J. Sharpton, M. M. Finucane, G. Jospin, S. W. Kembel, J. O'Dwyer, A. F. Koeppel, J. L. Green, and K. S. Pollard. (2013). "Global Marine Bacterial Diversity Peaks at High Latitudes in Winter." *ISME Journal* 7: e1669.

Lambshead, P.J.D., C. J. Brown, T. J. Ferrero, N. J. Mitchell, C. R. Smith, L. E. Hawkins, and J. Tietjen. (2002). "Latitudinal Diversity Patterns of Deep-sea Marine Nematodes and Organic Fluxes: A Test from the Central Equatorial Pacific." *Marine Ecology Progress Series* 236: 129–135.

Lambshead, P.J.D., J. Tietjen, T. Ferrero, and P. Jensen. (2000). "Latitudinal Diversity Gradients in the Deep Sea with Special Feference to North Atlantic Nematode." *Marine Ecology Progress Series* 194: 159–167.

Lavigne, D., S. Innes, G. Worthy, K. Kovacs, O. Schmitz, and J. Hickie. (1986). "Metabolic Rates of Seals and Whales." *Canadian Journal of Zoology* 64, 279–284.

Leibold, M. A., and M. A. McPeek. (2006). "Coexistence of the Niche and Neutral Perspective in Community Ecology." *Ecology* 87: 1399–1410.

Leprieur, F., P. Descombes, T. Gaboriau, P. F. Cowman, V. Parravicin, M. Kulbicki, C. J. Melian, C.N.D. Santana, C. Heine, D. Mouillot, D. R. Bellwood, and L. Pellissie. (2016). "Plate Tectonics Drive Tropical Reef Biodiversity Dynamics." *Nature Communications* 7: e11461.

Levin, L. A., and N. Le Bris. (2015). "The Deep Ocean under Climate Change." *Science* 350: 766–768.

Lewandowska, A. M., D. G. Boyce, M. Hofmann, B. Matthiessen, U. Sommer, and B. Worm. (2014). "Effects of Sea Surface Warming on Marine Plankton." *Ecology Letters* 17, 614–623.

Lieth, H., and R. H. Whittaker. (2012). *Primary Productivity of the Biosphere*. New York: Springer.

Lindgren, D.A.G. (1972). "The Temperature Influence on the Spontaneous Mutation Rate." *Hereditas* 70: 165–177.

Linnæus, C. (1758). *Systema naturæ per regna tria naturæ, secundum classes, ordines, genera, species, cum characteribus, differentiis, synonymis, locis*. Stockholm: L. Salvii.

Lotze, H. K., H. S. Lenihan, B. J. Bourque, R. Bradbury, R. G. Cooke, M. C. Kay, S. M. Kidwell, M. X. Kirby, C. H. Peterson, and J.B.C. Jackson. (2006). "Depletion, Degradation, and Recovery Potential of Estuaries and Coastal Seas." *Science* 312: 1806–1809.

Lotze, H. K., and L. McClenachan. (2013). "Marine Historical Ecology." In *Marine Community Ecology and Conservation*, ed. M. D. Bertness. New York: Sinauer, 165–200.

Lucifora, L., V. García, and B. Worm. (2011). "Global Diversity Hotspots and Conservation Priorities for Sharks." *PLoS ONE* 6: e19356.

MacArthur, R. H., and E. O. Wilson. (1967). *The Theory of Island Biography*. Princeton, NJ: Princeton University Press.

Macpherson, E. (2002). "Large-scale Species-richness Gradients in the Atlantic Ocean." *Proceedings of the Royal Society B; Biological Sciences* 269: 1715–1720.

Macpherson, E., P. A. Hastings, and D. R. Robertson. (2009). "Macroecological Patterns among Marine Fishes." In *Marine Macroecology*, ed. J. D. Witman and K. Roy. Chicago: University of Chicago Press, 122–152.

Malviya, S., E. Scalco, S. Audic, F. Vincent, A. Veluchamy, J. Poulain, P. Wincker, D. Iudicone, C. de Vargas, L. Bittner, A. Zingone, and C. Bowler. (2016). "Insights into Global Diatom Distribution and Diversity in the World's Ocean." *Proceedings of the National Academy of Sciences* 113: E1516–E1525.

Mangel, M. (2002). "The Important Role of Theory in Conservation Biology." *Conservation Biology* 16: 843–844.

Mannion, P. D., R. B. Benson, P. Upchurch, R. J. Butler, M. T. Carrano, and P. M. Barrett. (2012). "A Temperate Palaeodiversity Peak in Mesozoic Dinosaurs and Evidence for Late Cretaceous Geographical Partitioning." *Global Ecology and Biogeography* 21: 898–908.

Mannion, P. D., P. Upchurch, R. B. Benson, and A. Goswami. (2014). "The Latitudinal Biodiversity Gradient through Deep Time." *Trends in Ecology and Evolution* 29: 42–50.

Margules, C. R., and R. L. Pressey. (2000). "Systematic Conservation Planning." *Nature* 405: 243–253.

Marquet, P. A., A. P. Allen, J. H. Brown, J. A. Dunne, B. J. Enquist, J. F. Gillooly, P. A. Gowaty, J. L. Green, J. Harte, S. P. Hubbell, J. O'Dwyer et al. (2014). "On Theory in Ecology." *BioScience* 64: 701–710.

Martin, A. P., and S. R. Palumbi. (1993). "Body Size, Metabolic Rate, Generation Time, and the Molecular Clock." *Proceedings of the National Academy of Sciences* 90, 4087–4091.

Martin, W., J. Baross, D. Kelley, and M. J. Russell. (2008). "Hydrothermal Vents and the Origin of Life." *Nature Reviews Microbiology* 6: 805–814.

Matsuba, C., D. G. Ostrow, M. P. Salomon, A. Tolani, and C. F. Baer. (2013). "Temperature, Stress and Spontaneous Mutation in *Caenorhabditis briggsae* and *Caenorhabditis elegans*." *Biology Letters* 9: e20120334.

May, R. M. (1988). "How Many Species Are There on Earth?" *Science* 241: 1441–1449.

———. (1992). "Bottoms Up for the Oceans." *Nature* 357: 278–279.

Mayhew, P. J., M. A. Bell, T. G. Benton, and A. J. McGowan. (2012). "Biodiversity Tracks Temperature over Time." *Proceedings of the National Academy of Sciences USA* 109: 15141–15145.

Mayhew, P. J., G. B. Jenkins, and T. G. Benton. (2007). "A Long-term Association between Global Temperature and Biodiversity, Origination and Extinction in the Fossil Record." *Proceedings of the Royal Society B: Biological Sciences* 275: 47–53.

McCauley, D. J., M. L. Pinsky, S. R. Palumbi, J. A. Estes, F. H. Joyce, and R. R. Warner. (2015). "Marine Defaunation: Animal Loss in the Global Ocean." *Science* 347: c1255641.

McClain, C. R., A. P. Allen, D. P. Tittensor, and M. A. Rex. (2012). "Energetics of Life on the Deep Seafloor." *Proceedings of the National Academy of Sciences* 109: 15366–15371.

McClanahan, T., N. Muthiga, and S. Mangi. (2001). "Coral and Algal Changes after the 1998 Coral Bleaching." *Coral Reefs* 19: 380–391.

McCoy, E. D., and K. L. Heck. (1975). Biogeography of Corals, Mangroves, and Seagrasses: An Alternative to the Center of Origin Concept." *Systematic Zoology* 25: 201–210.

McGill, B. J., B. A. Maurer, and M. D. Weiser. (2006). "Empirical Evaluation of Neutral Theory." *Ecology* 87: 1411–1423.

Menéndez, R., A. G. Megías, J. K. Hill, B. Braschler, S. G. Willis, Y. Collingham, R. Fox, D. B. Roy, and C. D. Thomas. (2006). "Species Richness Changes Lag behind Climate Change." *Proceedings of the Royal Society B: Biological Sciences* 273: 1465–1470.

Michel, C., B. Bluhm, V. Gallucci, A. J. Gaston, F.J.L. Gordillo, R. Gradinger, R. Hopcroft, N. Jensen, T. Mustonen, A. Niemi, and T. G. Nielsen. (2012). "Biodiversity of Arctic Marine Ecosystems and Responses to Climate Change." *Biodiversity* 13: 200–214.

Micheli, F., and B. S. Halpern. (2005). "Low Functional Redundancy in Coastal Marine Assemblages." *Ecology Letters* 8: 391–400.

Miller, S. L., and L. E. Orgel. (1974). *The Origins of Life on the Earth*. Englewood Cliffs, NJ: Prentice-Hall.

Milner-Gulland, E. J., E. L. Bennett, and SCB 2002 Annual Meeting Wild Meat Group. (2003). "Wild Meat: The Bigger Picture." *Trends in Ecology and Evolution* 18: 351–357.

Mittelbach, G. G., C. F. Steiner, S. M. Scheiner, K. L. Gross, L. H. Reynolds, R. B. Waide, M. R. Willig, S. I. Dodson, and L. Gough. (2001). "What Is the Observed Relationship between Species Richness and Productivity?" *Ecology* 82: 2381–2396.

Molinos, J. G., B. S. Halpern, D. S. Schoeman, C. J. Brown, W. Kiessling, P. J. Moore, J. M. Pandolfi, E. S. Poloczanska, A. J. Richardson, and M. T. Burrows. (2016). "Climate Velocity and the Future Global Redistribution of Marine Biodiversity." *Nature Climate Change* 6: 83–88.

Mongold, J. A., A. F. Bennett, and R. E. Lenski. (1996). "Evolutionary Adaptation to Temperature. IV. Adaptation of *Escherichia Coli* at a Niche Boundary." *Evolution* 35–43.

Mora, C., A. G. Frazier, R. J. Longman, R. S. Dacks, M. M. Walton, E. J. Tong, J. J. Sanchez, L. R. Kaiser, Y. O. Stender, J. M. Anderson, C. M. Ambrosino et al. (2013). "The Projected Timing of Climate Departure from Recent Variability." *Nature* 502: 183–187.

Mora, C., D. P. Tittensor, S. Adl, A.G.B. Simpson, and B. Worm. (2011). "How Many Species Are There on Earth and in the Ocean?" *PLoS Biology* 9: e1001127.

Mora, C., D. P. Tittensor, and R. A. Myers. (2008). "The Completeness of Taxonomic Inventories for Describing the Global Diversity and Distribution of Marine Fishes." *Proceedings of the Royal Society B: Biological Sciences* 275: 149–155.

Mulkidjanian, A. Y., A. Y. Bychkov, D. V. Dibrova, M. Y. Galperin, and E. V. Koonin. (2012). "Origin of First Cells at Terrestrial, Anoxic Geothermal Fields." *Proceedings of the National Academy of Sciences* 109: E821–E830.

Muller, H. (1928). "The Measurement of Gene Mutation Rate in Drosophila, Its High Variability, and Its Dependence upon Temperature." *Genetics* 13: 279–357.

Mumby, P. J., A. J. Edwards, J. E. Arias-González, K. C. Lindeman, P. G. Blackwell, A. Gall, M. I. Gorczynska, A. R. Harborne, C. L. Pescod, H. Renken, C.C.C. Wabnitz, and G. Llewellyn. (2004). "Mangroves Enhance the Biomass of Coral Reef Fish Communities in the Caribbean." *Nature* 427: 533–536.

Myers, N., R. A. Mittermeier, C. G. Mittermeier, G.A.B. da Fonseca, and J. Kent. (2000). "Biodiversity Hotspots for Conservation Priorities." *Nature* 403: 853–858.

Nemani, R. R., C. D. Keeling, H. Hashimoto, W. M. Jolly, S. C. Piper, C. J. Tucker, R. B. Myneni, and S. W. Running. (2003). "Climate-driven Increases in Global Terrestrial Net Primary Production from 1982 to 1999." *Science* 300: 1560–1563.

Newbold, T., L. N. Hudson, S. L. Hill, S. Contu, I. Lysenko, R. A. Senior, L. Börger, D. J. Bennett, A. Choimes, and B. Collen. (2015). "Global Effects of Land Use on Local Terrestrial Biodiversity." *Nature* 520: 45–50.

Olson, D. B., G. L. Hitchcock, A. J. Mariano, C. J. Ashjan, G. Peng, R. W. Nero, and G. P. Podesta. (1994). "Life on the Edge: Marine Life and Fronts." *Oceanography* 7: 52–60.

Öpik, H., and S. A. Rolfe. (2005). *The Physiology of Flowering Plants*. Cambridge, UK: Cambridge University Press.

Orme, C.D.L., R. G. Davies, M. Burgess, F. Eigenbrod, N. Pickup, V. A. Olson, A. J. Webster, T.-S. Ding, P. C. Rasmussen, R. S. Ridgely, A. J. Stattersfield et al. (2005). "Global Hotspots of Species Richness Are Not Congruent with Endemism or Threat." *Nature* 436: 1016–1019.

Paine, R. T. (1966). "Food Web Complexity and Species Diversity." *American Naturalist* 100: 65–76.

———. (1984). "Ecological Determinism in the Competition for Space." *Ecology* 65: 1339–1348.

Palumbi, S. R., D. J. Barshis, N. Traylor-Knowles, and R. A. Bay. (2014). "Mechanisms of Reef Coral Resistance to Future Climate Change." *Science* 344: 895–898.

Pandolfi, J. M., R. H. Bradbury, E. Sala, T. P. Hughes, K. A. Bjorndal, R. G. Cooke, D. McArdle, L. McClenachan, M.J.H. Newman, G. Paredes, R. R. Warner, and J.B.C. Jackson. (2003). "Global Trajectories of the Long-term Decline of Coral Reef Ecosystems." *Science* 301: 955–958.

Parmesan, C. (2006). "Ecological and Evolutionary Responses to Recent Climate Change." *Annual Reviews in Ecology and Systematics* 37: 637–669.

Parmesan, C., and G. Yohe. (2003). "A Globally Coherent Fingerprint of Climate Change Impacts across Natural Systems." *Nature* 421: 37–42.

Pauly, D. (1995). "Anecdotes and the Shifting Baseline Syndrome of Fisheries." *Trends in Ecology and Evolution* 10: 430.

Pauly, D., J. Alder, E. Bennett, V. Christensen, P. Tyedmers, and R. Watson. (2003). "The Future for Fisheries." *Science* 302: 1359–1361.

Pauly, D., V. Christensen, and C. Walters. (2000). "Ecopath, Ecosim, and Ecospace as Tools for Evaluating Ecosystem Impacts of Fisheries." *ICES Journal of Marine Science* 57: 697–706.

Pauly, D., and D. Zeller D. (2016). "Catch Reconstructions Reveal That Global Marine Fisheries Catches Are Higher than Reported and Declining." *Nature Communications* 7: e10244.

Pearman, P. B., A. Guisan, O. Broennimann, and C. F. Randin. (2008). "Niche Dynamics in Space and Time." *Trends in Ecology and Evolution* 23: 149–158.

Peet, R. K. (1974). "The Measurement of Species Diversity." *Annual Review of Ecology and Systematics* 5: 285–307.

Perry, A. L., Low P. J., Ellis J. R., and Reynolds J. D. (2005). "Climate Change and Distribution Shifts in Marine Fishes." *Science* 308: 1912–1915.

Peterson, A., J. Soberón, and V. Sánchez-Cordero. (1999). "Conservatism of Ecological Niches in Evolutionary Time." *Science* 285: 1265–1267.

Petraitis, P. S., R. E. Latham, and R. A. Niesenbaum. (1989). "The Maintenance of Species Diversity by Disturbance." *Quarterly Reviews in Biology* 64: 393–418.

Pianka, E. R. (1966). "Latitudinal Gradients in Species Diversity: A Review of Concepts." *American Naturalist* 100: 33–46.

Pickett, S.T.A., and P. S. White. (1985). *The Ecology of Natural Disturbance and Patch Dynamics*. New York: Academic Press.

Pielou, E. C. (1979). *Biogeography*. New York: Wiley.

Pimm, S. L., G. J. Russell, J. L. Gittleman, and T. M. Brooks. (1995). "The Future of Biodiversity." *Science* 269: 347–350.

Pinsky, M. L., B. Worm, M. J. Fogarty, J. L. Sarmiento, and S. A. Levin. (2013). "Marine Taxa Track Local Climate Velocities." *Science* 341: 1239–1242.

Polidoro, B. A., K. E. Carpenter, L. Collins, N. C. Duke, A. M. Ellison, J. C. Ellison, E. J. Farnsworth, E. S. Fernando, K. Kathiresan, N. E. Koedam, S. R. Livingstone et al. (2010). "The Loss of Species: Mangrove Extinction Risk and Geographic Areas of Global Concern." *PLoS ONE* 5: e10095.

Polovina, J. J., E. A. Howell, and M. Abecassis. (2008). "Ocean's Least Productive Waters Are Expanding." *Geophysical Research Letters* 35: L03618.

Polovina, J. J., E. Howell, D. R. Kobayashi, and M. P. Seki. (2001). "The Transition Zone Chlorophyll Front, a Dynamic Global Feature Defining Migration and Forage Habitat for Marine Resources." *Progress in Oceanography* 49: 469–483.

Pommier, T., B. Canbäck, L. Riemann, K. H. Boström, K. Simu, P. Lundberg, A. Tunlid, and Å Hagström. (2007). "Global Patterns of Diversity and Community Structure in Marine Bacterioplankton." *Molecular Ecology* 16: 867–880.

Pounds, J. A., M. R. Bustamante, L. A. Coloma, J. A. Consuegra, M.P.L. Fogden, P. N. Foster, E. L. Marca, K. L. Masters, A. Merino-Viteri, R. Puschendorf, S. R. Ron et al. (2006). "Widespread Amphibian Extinctions from Epidemic Disease Driven by Global Warming." *Nature* 437: 161–167.

Powell, M. G., V. P. Beresford, and B. A. Colaianne. (2012). "The Latitudinal Position of Peak Marine Diversity in Living and Fossil Biotas." *Journal of Biogeography* 39: 1687–1694.

Pressey, R. L., M. Cabeza, M. E. Watts, R. M. Cowling, and K. A. Wilson. (2007). "Conservation Planning in a Changing World." *Trends in Ecology and Evolution* 22: 583–592.

Preston, F. W. (1962). "The Canonical Distribution of Commonness and Rarity: Part I." *Ecology* 43: 185–215.

Price, P. W., G. W. Fernandes, A.C.F. Lara, J. Brawn, H. Barrios, M. G. Wright, S. P. Ribeiro, and N. Rothcliff. (1998). "Global Patterns in Local Number of Insect Galling Species." *Journal of Biogeography* 25: 581–591.

Purves, D., J.P.W. Scharlemann, M. Harfoot, T. Newbold, D. P. Tittensor, J. Hutton, and S. Emmott. (2013). "Time to Model All Life on Earth." *Nature* 493: 295–297.

Raes, J., I. Letunic, T. Yamada, L. J. Jensen, and P. Bork. (2011). "Toward Molecular Trait-based Ecology through Integration of Biogeochemical, Geographical and Metagenomic Data." *Molecular Systems Biology* 7: 473–474.

Ramirez-Llodra, E., P. A. Tyler, M. C. Baker, O. A. Bergstad, M. R. Clark, E. Escobar, L. A. Levin, L. Menot, A. A. Rowden, and C. R. Smith. (2011). "Man and the Last Great Wilderness: Human Impact on the Deep Sea." *PLoS ONE* 6: e22588.

Raup, D. M., and J.J.J. Sepkoski. (1982). "Mass Extinctions in the Marine Fossil Record." *Science* 215: 1501–1503.

Renema, W., D. R. Bellwood, J. C. Braga, K. Bromfield, R. Hall, K. G. Johnson, P. Lunt, C. P. Meyer, L. B. McMonagle, R. J. Morley, A. O'Dea et al. (2008). "Hopping Hotspots: Global Shifts in Marine Biodiversity." *Science* 321: 654–657.

Retallack, G. J. (2013). "Ediacaran Life on Land." *Nature* 493: 89–92.

Rex, M. A., and R. J. Etter. (2010). *Deep-sea Biodiversity: Pattern and Scale.* Cambridge, MA: Harvard University Press.

Rex, M. A., C. T. Stuart, and R. J. Etter. (2001). "Do Deep-sea Nematodes Show a Positive Latitudinal Gradient of Species Diversity? The Potential Role of Depth." *Marine Ecology Progress Series* 210: 297–298.

Rex, M. A., C. T. Stuart, R. L. Hessler, J. A. Allen, H. L. Sanders, and G.D.F. Wilson. (1993). "Global-scale Latitudinal Patterns of Species Diversity in the Deep-sea Benthos." *Nature* 365: 636–639.

Reynolds, J. D., N. K. Dulvy, N. B. Goodwin, and J. A. Hutchings. (2005). "Biology of extinction risk in marine fishes." *Proceedings of the Royal Society B: Biological Sciences* 272: 2337–2344.

Richardson, A. J., and D. S. Schoeman. (2004). "Climate Impact on Plankton Ecosystems in the Northeast Atlantic." *Science* 305: 1609–1612.

Roberts, C. M. (2002). "Deep Impact: The Rising Toll of Fishing in the Deep Sea." *Trends in Ecology and Evolution* 242: 242–245.

Roberts, C. M., C. J. McClean, J.E.N. Veron, J. P. Hawkins, G. R. Allen, D. E. McAllister, C. G. Mittermeier, F. W. Schueler, M. Spalding, F. Wells, C. Vynne, and T. B. Werner. (2002). "Marine Biodiversity Hotspots and Conservation Priorities for Tropical Reefs." *Science* 295: 1280–1284.

Rockstrom, J., W. Steffen, K. Noone, A. Persson, F. S. Chapin, E. F. Lambin, T. M. Lenton, M. Scheffer, C. Folke, H. J. Schellnhuber, B. Nykvist et al. (2009). "A Safe Operating Space for Humanity." *Nature* 461: 472–475.

Rohde, K. (1992). "Latitudinal Gradients in Species-diversity—The Search for the Primary Cause." *Oikos* 65: 514–527.

Rombouts, I., G. Beaugrand, F. Ibañez, S. Gasparini, S. Chiba, and L. Legendre. (2009). "Global Latitudinal Variations in Marine Copepod Diversity and Environmental Factors." *Proceedings of the Royal Society B: Biological Sciences* 276: 3053–3062.

Root, T. L., J. T. Price, K. R. Hall, S. H. Schneider, C. Rosenzweig, and J. A. Pounds. (2003). "Fingerprints of Global Warming on Wild Animals and Plants." *Nature* 421: 57–60.

Roper, C.F.E., and E. K. Shea. (2013). "Unanswered Questions about the Giant Squid *Architeuthis* (Architeuthidae) Illustrate Our Incomplete Knowledge of Coleoid Cephalopods." *American Malacological Bulletin* 31: 109–122.

Rosenblatt, R. H. (1967). "The Zoogeographic Relationships of the Marine Shore Fishes of Tropical America." *Studies in Tropical Ecology* 5: 579–592.

Rosenzweig, M. L., and Z. Abramsky. (1993). "How Are Diversity and Productivity Related?" In *Species Diversity in Ecological Communities*, ed. R. E. Rickleffs and D. Schluter. Chicago: University of Chicago Press, 52–65.

Rosindell, J., S. J. Cornell, S. P. Hubbell, and R. S. Etienne. (2010). "Protracted Speciation Revitalizes the Neutral Theory of Biodiversity." *Ecology Letters* 13: 716–727.

Rosindell, J., S. P. Hubbell, and R. S. Etienne. (2011). "The Unified Neutral Theory of Biodiversity and Biogeography at Age Ten." *Trends in Ecology and Evolution* 26: 340–348.

Rosindell, J., S. P. Hubbell, F. He, L. J. Harmon, and R. S. Etienne. (2012). "The Case for Ecological Neutral Theory." *Trends in Ecology and Evolution* 27: 203–208.

Rosindell, J., Y. Wong, and R. S. Etienne. (2008). "A Coalescence Approach to Spatial Neutral Ecology." *Ecological Informatics* 3: 259–271.

Roy, K., and J. D. Witman. (2009). "Spatial Patterns of Species Diversity in the Shallow Marine Invertebrates: Patterns, Processes, and Prospects." In *Marine Macroecology*, ed. J. D. Witman and K. Roy. Chicago: University of Chicago Press, 101–121.

Ruddiman, W. F. (1969). "Recent Planktonic Foraminifera: Dominance and Diversity in North Atlantic Surface Sediments." *Science* 164: 1164–1167.

Rutherford, S., S. D'Hondt, and W. Prell. (1999). "Environmental Controls on the Geographic Distribution of Zooplankton Diversity." *Nature* 400: 749–753.

Safi, K., M. V. Cianciaruso, R. D. Loyola, D. Brito, K. Armour-Marshall, and J.A.F. Diniz-Filho. (2011). "Understanding Global Patterns of Mammalian Functional and Phylogenetic Diversity." *Philosophical Transactions of the Royal Society B: Biological Sciences* 366: 2536–2544.

Sala, O. E., F. S. Chapin III, J. J. Armesto, E. Berlow, J. Bloomfield, R. Dirzo, E. Huber-Sanwald, L. F. Huenneke, R. B. Jackson, A. Kinzig, R. Leemans et al. (2000). "Global Biodiversity Scenarios for the Year 2100." *Science* 287: 1770–1774.

Sanders, H. L. (1968). "Marine Benthic Diversity: A Comparative Study." *American Naturalist* 102: 243–282.

Sarmiento, J. L., R. Slater, R. Barber, L. Bopp, S. C. Doney, A. C. Hirst, J. Kleypas, R. Matear, U. Mikolajewicz, P. Monfray, V. Soldatov et al. (2004). "Response of Ocean Ecosystems to Climate Warming." *Global Biogeochemical Cycles* 18; doi:10.1029/2003GB002134.

Schipper, J., J. S. Chanson, F. Chiozza, N. A. Cox, M. Hoffmann, V. Katariya, J. Lamoreux, A.S.L. Rodrigues, S. N. Stuart, H. J. Temple, J. Baillie et al. (2008). "The Status of the World's Land and Marine Mammals: Diversity, Threat, and Knowledge." *Science* 322: 225–230.

Schlüter, L., K. T. Lohbeck, M. A. Gutowska, J. P. Gröger, U. Riebesell, and T. B. Reusch. (2014). "Adaptation of a Globally Important Coccolithophore to Ocean Warming and Acidification." *Nature Climate Change* 4: 1024–1030.

Selig, E. R., W. R. Turner, S. Troëng, B. P. Wallace, B. S. Halpern, K. Kaschner, B. G. Lascelles, K. E. Carpenter, and R. A. Mittermeier. (2014). "Global Priorities for Marine Biodiversity Conservation." *PLoS ONE* 9: e82898.

Sepkoski, J. J. (1998). "Rates of Speciation in the Fossil Record." *Philosophical Transactions of the Royal Society B: Biological Sciences* 353: 315–326.

Short, F. T., B. Polidoro, S. R. Livingstone, K. E. Carpenter, S. Bandeira, J. S. Bujang, H. P. Calumpong, T.J.B. Carruthers, R. G. Coles, W. C. Dennison, P.L.A. Erftemeijer et al. (2011). "Extinction Risk Assessment of the World's Seagrass Species." *Biological Conservation* 144, 1961–1971.

Silvertown, J., M. E. Dodd, K. McConway, J. Potts, and M. Crawley. (1994). "Rainfall, Biomass Variation, and Community Composition in the Park Grass Experiment." *Ecology* 75: 2430–2437.

Sinervo, B., F. Méndez-de-la-Cruz, D. B. Miles, B. Heulin, E. Bastiaans, M. Villagrán-Santa Cruz, R. Lara-Resendiz, N. Martínez-Méndez, M. L. Calderón-Espinosa, R. N. Meza-Lázaro, H. Gadsden et al. (2010). "Erosion of Lizard Diversity by Climate Change and Altered Thermal Niches." *Science* 328: 894–899.

Sommer, J. H., H. Kreft, G. Kier, W. Jetz, J. Mutke, and W. Barthlott. (2010). "Projected Impacts of Climate Change on Regional Capacities for Global Plant Species Richness." *Proceedings of the Royal Society of London B: Biological Sciences*; doi:10.1098/rspb.2010.0120.

Sommer, U., and B. Worm. (2002). *Competition and Coexistence*. Berlin: Springer.

Sousa, W. P. (1979). "Disturbance in Marine Intertidal Boulder Fields: The Nonequilibrium Maintenance of Species Diversity." *Ecology* 49: 1225–1239.

Stehli, F. G., R. G. Douglas, and N. D. Newell. (1969). "Generation and Maintenance of Gradients in Taxonomic Diversity." *Science* 164: 947–949.

Stehli, F. G., A. L. McAlester, and C. E. Helsley. (1967). "Taxonomic Diversity of Recent Bivalves and Some Implications for Geology." *Geological Society of America Bulletin* 78: 455–466.

Stehli, F. G., and J. W. Wells. (1971). "Diversity and Age Patterns in Hermatypic Corals." *Systematic Biology* 20: 115–126.

Stevens, C. J., N. B. Dise, J. O. Mountford, and D. J. Gowing. (2004). "Impact of Nitrogen Deposition on the Species Richness of Grasslands." *Science* 303: 1876–1879.

Stevens, G. C. (1989). "The Latitudinal Gradient in Geographical Range—How So Many Species Coexist in the Tropics." *American Naturalist* 133: 240–256.

Storch, D., P. A. Marquet, and J. H. Brown. (2007). *Scaling Biodiversity*. Cambridge, UK: Cambridge University Press.

Stramma, L., S. Schmidtko, L. A. Levin, and G. C. Johnson. (2010). "Ocean Oxygen Minima Expansions and Their Biological Impacts." *Deep-Sea Research Part One* 57: 587–595.

Stuart-Smith, R. D., A. E. Bates, J. S. Lefcheck, J. E. Duffy, S. C. Baker, R. J. Thomson, J. F. Stuart-Smith, N. A. Hill, S. J. Kininmonth, L. Airoldi, M. A. Becerro et al. (2013). "Integrating Abundance and Functional Traits Reveals New Global Hotspots of Fish Diversity." *Nature* 501: 539–542.

Stuart-Smith, R. D., G. J. Edgar, N. S. Barrett, S. J. Kininmonth, and A. E. Bates. (2015). "Thermal Biases and Vulnerability to Warming in the World's Marine Fauna." *Nature* 528: 88–92.

Sul, W. J., T. A. Oliver, H. W. Ducklow, L. A. Amaral-Zettler, and M. L. Sogin. (2013). "Marine Bacteria Exhibit a Bipolar Distribution." *Proceedings of the National Academy of Sciences USA* 110: 2342–2347.

Sun, Y., M. M. Joachimski, P. B. Wignall, C. Yan, Y. Chen, H. Jiang, L. Wang, and X. Lai. (2012). "Lethally Hot Temperatures during the Early Triassic Greenhouse." *Science* 338: 366–370.

Sunagawa, S., L. P. Coelho, S. Chaffron, J. R. Kultima, K. Labadie, G. Salazar, B. Djahanschiri, G. Zeller, D. R. Mende, and A. Alberti. (2015). "Structure and Function of the Global Ocean Microbiome." *Science* 348: e1261359.

Sunday, J. M., A. E. Bates, and N. K. Dulvy. (2011). "Global Analysis of Thermal Tolerance and Latitude in Ectotherms." *Proceedings of the Royal Society B: Biological Sciences* 278: 1823–1830.

Sunday, J. M., A. E. Bates, and N. K. Dulvy. (2012). "Thermal Tolerance and the Global Redistribution of Animals." *Nature Climate Change* 2: 686–690.

Sydeman, W. J., E. Poloczanska, T. E. Reed, and S. A. Thompson. (2015). "Climate Change and Marine Vertebrates." *Science* 350: 772–777.

Tashiro, T., A. Ishida, M. Hori, M. Igisu, M. Koike, P. Méjean, N. Takahata, Y. Sano, and T. Komiya. (2017). "Early Trace of Life from 3.95 Ga Sedimentary Rocks in Labrador, Canada." *Nature* 549: 516–518.

Taylor, J., and C. Taylor. (1977). "Latitudinal Distribution of Predatory Gastropods on the Eastern Atlantic Shelf." *Journal of Biogeography* 4: 73–81.

Tedersoo, L., M. Bahram, S. Põlme, U. Kõljalg, N. S. Yorou, R. Wijesundera, L. V. Ruiz, A. M. Vasco-Palacios, P. Q. Thu, A. Suija, M. E. Smith et al. (2014). "Global Diversity and Geography of Soil Fungi." *Science* 346: e1256688.

Thiery, R. G. (1982). "Environmental Instability and Community Diversity." *Biological Reviews* 57: 691–710.

Thomas, C. D., A. Cameron, R. E. Green, M. Bakkenes, L. J. Beaumont, Y. C. Collingham, B.F.N. Erasmus, M. F. de Siqueira, A. Grainger, L. Hannah, L. Hughes et al. (2004). "Extinction Risk from Climate Change." *Nature* 427: 145–148.

Thomas, M. K., C. T. Kremer, C. A. Klausmeier, and E. Litchman. (2012). "A Global Pattern of Thermal Adaptation in Marine Phytoplankton." *Science* 338: 1085–1088.

Thorson, G. (1952). "Zur jetzigen Lage der marinen Bodentier-Ökologie." *Zoologischer Anzeiger* 16: 276–327.

———. (1957). "Bottom Communities (Sublittoral or Shallow Shelf)." *Treatise on Marine Ecology and Paleoecology* 1: 461–534.

Tickell, C. (1997). "The Value of Diversity." In *Marine Biodiversity*, ed. R.F.G. Ormond, J. D. Gage, and M. V. Angel. Cambridge, UK: Cambridge University Press, xiii–xxii.

Tilman, D. (1982). *Resource Competition and Community Structure*. Princeton, NJ: Princeton University Press.

Tilman, D., R. M. May, C. L. Lehman, and M. A. Nowak. (1994). "Habitat Destruction and the Extinction Debt." *Nature* 371: 65–66.

Tisseuil, C., J.-F. Cornu, O. Beauchard, S. Brosse, W. Darwall, R. Holland, B. Hugueny, P. A. Tedesco, and T. Oberdorff. (2013). "Global Diversity Patterns and Cross-taxa Convergence in Freshwater Systems." *Journal of Animal Ecology* 82: 365–376.

Tittensor, D. P., C. Mora, W. Jetz, H. K. Lotze, D. Ricard, E. V. Berghe, and B. Worm. (2010). "Global Patterns and Predictors of Marine Biodiversity across Taxa." *Nature* 466: 1098–1101.

Tittensor, D. P., M. A. Rex, C. T. Stuart, C. R. McClain, and C. R. Smith. (2011). "Species–Energy Relationships in Deep-sea Molluscs." *Biology Letters* 7: e20101174.

Tittensor, D. P., M. Walpole, S.L.L. Hill, D. G. Boyce, G. L. Britten, N. D. Burgess, S.H.M. Butchart, P. W. LeadleyE. C. Regan, R. Alkemade, R. Baumung et al. (2014). "A Mid-term Analysis of Progress toward International Biodiversity Targets." *Science* 346: 241–244.

Tittensor, D. P., and B. Worm. (2016). "A Neutral-Metabolic Theory of Latitudinal Biodiversity." *Global Ecology and Biogeography* 25: 630–641.

UN (1992). *Convention on Biological Diversity*. New York: United Nations. Available at http://www. biodiv.org/.

Urban, M. C. (2015). "Accelerating Extinction Risk from Climate Change." *Science* 348: 571–573.

Valentine, J. W., and D. Jablonski. (2015). "A Twofold Role for Global Energy Gradients in Marine Biodiversity Trends." *Journal of Biogeography* 42: 997–1005.

van Herk, C. M., A. Aptroot, and H. F. van Dobben. (2002). "Long-term Monitoring in the Netherlands Suggests That Lichens Respond to Global Warming." *Lichenologist* 34: 141–154.

Vasconcelos, R. P., S. Henriques, S. França, S. Pasquaud, I. Cardoso, M. Laborde, and H. N. Cabral. (2015). "Global Patterns and Predictors of Fish Species Richness in Estuaries." *Journal of Animal Ecology* 84: 1331–1341.

Vellend, M. (2010). "Conceptual Synthesis in Community Ecology." *Quarterly Review of Biology* 85: 183–206.

———. (2016). *The Theory of Ecological Communities*. Princeton, NJ: Princeton University Press.

Vitousek, P. M. (1994). "Beyond Global Warming: Ecology and Global Change." *Ecology* 75: 1861–1876.

Vitousek, P. M., H. A. Mooney, J. Lubchenco, and J. M. Melillo. (1997). "Human Domina-
tion of Earth's Ecosystems." *Science* 277: 494–499.

Walther, G.-R., E. Post, P. Convey, A. Menzel, C. Parmesan, T.J.C. Beebee, J.-M. Fromen-
tin, O. Hoegh-Guldberg, and F. Bairlein. (2002). "Ecological Responses to Recent
Climate Change." *Nature* 416: 351–460.

Wassmann, P., C. M. Duarte, S. Agusti, and M. K. Sejr. (2011). "Footprints of Climate
Change in the Arctic Marine Ecosystem." *Global Change Biology* 17: 1235–
1249.

Waters, C. N., J. Zalasiewicz, C. Summerhayes, A. D., Barnosky, C. Poirier, A. Gałuszka,
A. Cearreta, M. Edgeworth, E. C. Ellis, M. Ellis, C. Jeandel et al. (2016). "The
Anthropocene Is Functionally and Stratigraphically Distinct from the Holocene."
Science 351; doi:10.1126/science.aad2622.

Watling, L., and E. A. Norse. (1998). "Disturbance of the Seabed by Mobile Fishing Gear:
A Comparison to Forest Clearcutting." *Conservation Biology* 12: 1180–1197.

Webb, T. J. (2012). "Marine and Terrestrial Ecology: Unifying Concepts, Revealing Differences." *Trends in Ecology and Evolution* 27: 535–541.

Weigelt, P., M. J. Steinbauer, J. S. Cabral, and H. Kreft. (2016). "Late Quaternary Climate
Change Shapes Island Biodiversity." *Nature* 532: 99–102.

Wennekes, P., J. Rosindell, and R. Etienne. (2012). "The Neutral–Niche Debate: A Philosophical Perspective." *Acta Biotheoretica* 60: 257–271.

Węsławski, J. M., M. A. Kendall, M. Włodarska-Kowalczuk, K. Iken, M. Kędra,
J. Legezynska, and M. K. Sejr. (2011). "Climate Change Effects on Arctic Fjord
and Coastal Macrobenthic Diversity Observations and Predictions." *Marine Biodiversity* 41: 71–85.

West, G. B., J. H. Brown, and B. J. Enquist. (1997). "A General Model for the Origin of
Allometric Scaling Laws in Biology." *Science* 276: 122–126.

Whitehead, H., B. McGill, and B. Worm. (2008). "Diversity of Deep-water Cetaceans in
Relation to Temperature: Implications for Ocean Warming." *Ecology Letters* 11,
1198–1207.

Whittaker, R. H. (1975). *Communities and Ecosystems*. New York: Macmillan.

Wiens, J. J., and C. H. Graham. (2005). "Niche Conservatism: Integrating Evolution, Ecology, and Conservation Biology." Annual Reviews in Ecology and Systematics 36:
519–539.

Wildlife Conservation Society (WCS), and Center for International Earth Science Information Network (CIESIN), Columbia University. (2005). Last of the Wild Project, Version 2, 2005 (LWP-2): Global Human Footprint Dataset (Geographic). Palisades,
NY: NASA Socioeconomic Data and Applications Center (SEDAC). http://dx.doi
.org/10.7927/H4M61H5F. Accessed November 14, 2017.

Wilson, S. K., N.A.J. Graham, M. S. Pratchett, G. P. Jones, and N.V.C. Polunin. (2006).
"Multiple Disturbances and the Global Degradation of Coral Reefs: Are Reef Fishes
at Risk or Resilient?" *Global Change Biology* 12: 2220–2234.

Woolley, S.N.C., D. P. Tittensor, G. Guillera-Arroita, J. J. Lahoz-Monfort, B. A. Wintle,
B. Worm, and T. D. O'Hara. (2016). "Deep-sea Diversity Patterns Shaped by Energy
Availability." *Nature* 533: 393–396.

Worm, B., E. B. Barbier, N. Beaumont, J. E. Duffy, C. Folke, B. S. Halpern, J.B.C.
Jackson, H. K. Lotze, F. Micheli, S. R. Palumbi, E. Sala et al. (2006). "Impacts of
Biodiversity Loss on Ocean Ecosystem Services." *Science* 314: 787–790.

Worm, B., and H. S. Lenihan. (2013). "Threats to Marine Ecosystems: Overfishing and Habitat Degradation." In *Marine Community Ecology and Conservation*, ed. M. D. Bertness. New York: Sinauer, 449–476.

Worm, B., H. K. Lotze, H. Hillebrand, and U. Sommer. (2002). "Consumer versus Resource Control of Species Diversity and Ecosystem Functioning." *Nature* 417: 848–851.

Worm, B., H. K. Lotze, I. Jonsen, and C. Muir. (2010). "The Future of Marine Animal Populations." In *Life in the World's Oceans: Diversity, Distribution and Abundance*, ed. A. McIntyre. Oxford, UK: Blackwell, 315–330.

Worm, B., H. K. Lotze, and R. A. Myers. (2003). "Predator Diversity Hotspots in the Blue Ocean." *Proceedings of the National Academy of Sciences USA* 100: 9884–9888.

Worm, B., and H. K. Lotze. (2015). "Marine Biodiversity and Climate Change." In Climate Change, ed. T. Letcher. Amsterdam: Elsevier, 195–212.

Worm, B., M. Sandow, A. Oschlies, H. K. Lotze, and R. A. Myers. (2005). "Global Patterns of Predator Diversity in the Open Oceans." *Science* 309: 1365–1369.

Worm, B., and D. P. Tittensor. (2011). "Range Contraction in Large Pelagic Predators." *Proceedings of the National Academy of Sciences* 108: 11942–11947.

Wright, D. H. (1983). "Species-Energy Theory: An Extension of Species-Area Theory." *Oikos* 41: 496–506.

Yasuhara, M., G. Hunt, H. J. Dowsett, M. M. Robinson, and D. K. Stoll. (2012). "Latitudinal Species Diversity Gradient of Marine Zooplankton for the Last Three Million Years." *Ecology Letters* 15: 1174–1179.

Yasuhara, M., D. P. Tittensor, H. Hillebrand, and B. Worm. (2015). "Combining Marine Macroecology and Palaeoecology in Understanding Biodiversity: Microfossils as a Model." *Biological Reviews*; 92: 199–215.

Zachos, J. C., G. R. Dickens, and R. E. Zeebe. (2008). "An Early Cenozoic Perspective on Greenhouse Warming and Carbon-cycle Dynamics." *Nature* 451: 279–283.

Zapata, F. A., K. J. Gaston, and S. L. Chown. (2005). "The Mid-domain Effect Revisited." *American Naturalist* 166: E144–E148.

Index

squids: environmental predictors of species richness, 78; global biodiversity data fit, 129; species richness of, 26, 132; synthesis of global biodiversity patterns, 52; univariate relationship of species richness and sea surface temperature, 84

stability-time hypothesis, 6, 65–66

surface ocean: environmental predictors of species richness in, 80

surface temperature: driver promoting diversity, 60–64

Systems Naturae (Linnaeus), 5

taxonomic resolution: sensitivity of vertebrate diversity patterns to, 45

temperature: biodiversity changes by, 157–59; community size and, 173–74; global impact on biodiversity, 156

temperature niches: dispersal rates of species, 114–16, 118; effects on individual species, 116, 120; effects on latitudinal ranges, 114–16; including in predicting global biodiversity, 143, 144–45; in model development, 113–21; niche width, 113, 115, 116, 117, 118; phytoplankton species, 113. *See also* neutral-metabolic-niche model; thermal niches

terrestrial biodiversity, 35–41; changes through time, 43; freshwater species, 40; in habitat, 4; hotspots of species richness across taxa, 154; land plants, 37; land vertebrates, 37–40; nutrient utilization in, 36; predicting, 135, 137, 138; sensitivity of predictions for, 139; synthesis of, 40–41

terrestrial birds: species richness of, 38, 39–40

theory development, 93–94; basic model implementation in, 97; basic neutral theory, 94–101; discussion and comparison with other theory, 122–23; including habitat area and productivity, 109, 111; including metabolic theory, 104–5, 107, 109; including temperature niches, 113–16, 118, 120–21; model implementation in forward and coalescence mode, 101–4; modeling toolbox, 174. *See also* neutral-metabolic-niche model

Theory of Island Biogeography (MacArthur and Wilson), 6, 8, 65, 71, 72, 94, 109, 149, 174

thermal energy: driver promoting diversity, 58, 60–64

thermal niche, 159, 175; conservatism of, 67, 68; global diversity predictions including, 143, 144–45, 147; global warming and, 157; hypothesis, 69, 82; model framework including, 113–16, 118, 120–21; seasonality and, 70; species distribution models and, 163; species richness patterns and, 176–77. *See also* temperature niches

Tilman, David, 1

time: biodiversity hotspots changing through, 42; changes in biodiversity patterns through, 41–44; ecological vs. evolutionary, 178–79; environmental drives of drivers over, 63

tuna and billfish: environmental predictors of species richness, 78; exploitation of, 161; global biodiversity data fit, 129; species richness of, 28, 30, 132; synthesis of global biodiversity patterns, 52; univariate relationship of species richness and sea surface temperature, 84

United Nations Sustainable Development Goals, 170

The Unified Neutral Theory of Biodiversity and Biogeography (Hubbell), 7, 149, 174

vascular plants: environmental predictors of species richness on land, 79; global biodiversity data fit, 129; mammals and, 182; synthesis of global biodiversity patterns, 53

vertebrates, 2; biodiversity of land, 37–40; sensitivity of biodiversity patterns to taxonomic resolution, 46–47. *See also* coastal vertebrates

von Humboldt, Alexander, 3, 5

white shark (*Carcharodon carcharias*), 21

zero-sum ecological dynamics, 97, 181; assumption of, 98, 176

zooplankton, 25, 160, 164

MONOGRAPHS IN POPULATION BIOLOGY

EDITED BY SIMON A. LEVIN AND HENRY S. HORN